The University
at the Crossroads to a
Sustainable Future

The Glion Colloquiums

Founded in 1998 by Luc Weber, University of Geneva, Werner Hirsch, UC Los Angeles, and James Duderstadt, University of Michigan, the Colloquium's objective is to allow leaders of renowned universities to meet and discuss various questions related to the development of science and Higher Education as well as governance and leadership of research intensive universities. The Colloquiums are organized by a small independent Association based in Geneva, Switzerland, and by an international programme Committee designated every other year to set up the program and invite participants. Various financial helps and funds have been found throughout the years. Research and cultural international foundations, global corporations, Swiss universities as well as the Swiss State Secretariat for education, research and innovation have participated.

Altogether, 150 different leading figures, active or recently retired university leaders, as well as some politicians and business leaders, have participated in one or more Colloquiums. Participants considered topics such as the rapidly changing nature of research universities, university governance, the interaction between universities and society, collaboration between universities and business, the globalization of higher education, and how universities prepare to address the changes characterizing our times. The contributions participants are invited to write beforehand openly reflect their views and experience in order to stimulate discussion. The Glion Colloquium sessions are held *in camera*, to guarantee open and genuine exchange.

To secure a dissemination as broad as possible of the analysis and recommendations coming out of the contributions and discussions, the revised contributions are published 6-8 months later in a volume which is freely distributed to numerous university leaders worldwide and sold commercially as well. This book is the twelves of the series. Nine of them have been published by ECONOMICA in Paris. Since the 11th book, the organizing Committee has opted for self-publication and a print-on-demand solution. Searchable PDFs of the books and of each of their composing chapters are freely available one year after publication on the Glion Colloquium's website (www.glion.org) and on the Open Archives of the University of Geneva (https://archive-ouverte.unige.ch/).

Volumes

1. *Challenges Facing Higher Education at the Millennium*, Werner Z. Hirsch and Luc E. Weber, eds, American Council on Education/Oryx Press, Phoenix and IAU Press/Pergamon, Paris and Oxford, (1999)
2. *Governance in Higher Education, The University in a State of Flux*, Werner Z. Hirsch and Luc E. Weber, eds, Economica, Paris, London, Geneva (2001)
3. *As the Walls of Academia are Tumbling Down*, Werner Z. Hirsch and Luc E. Weber, eds, Economica, Paris, London, Geneva (2002)
4. *Reinventing the Research University*, Luc E. Weber and James J. Duderstadt, eds, Economica, Paris, London, Geneva (2004)
5. *Universities and Business: Partnering for the Knowledge Economy*, Luc E. Weber and James J. Duderstadt, eds, Economica, Paris, London, Geneva (2006)
6. *The Globalization of Higher Education*, Luc E. Weber and James J. Duderstadt, eds, Economica, Paris, London, Geneva (2008)
7. *University Research for Innovation*, Luc E. Weber and James J. Duderstadt, eds, Economica, Paris, London, Geneva (2010)
8. *Global Sustainability and the Responsibilities of Universities*, Luc E. Weber and James J. Duderstadt, eds, Economica, Paris, London, Geneva (2012)
9. *Preparing Universities for an Era of Change*, Luc E. Weber and James J. Duderstadt, eds, Economica, Paris, London, Geneva (2014)
10. *University Priorities and Constraints*, Luc E. Weber and James J. Duderstadt, eds, Economica, Paris, London, Geneva (2016)
11. *The Future of the University in a Polarizing World*, Luc E. Weber and Howard Newby, eds, The Glion Colloquium, Geneva (2018)
12. *The University at the Crossroads to a Sustainable Future*, Luc E. Weber and Bert van der Zwaan, eds, The Glion Colloquium, Geneva (2020)

Declarations

1. Rhodes, F. H. T. *The First Glion Declaration: The University at the Millennium*, The Glion Colloquium (1998)
2. Rhodes, F. H. T. *The Second Glion Declaration: Universities and the Innovation Spirit*, The Glion Colloquium (2009)

The University
at the Crossroads to a
Sustainable Future

Edited by

Luc E. Weber

Bert van der Zwaan

Volume 12

Geneva

Published by the Association Glion Colloquium
c/o Rectorate, University of Geneva
24, rue Général Dufour
CH-1211 Geneva 4
Switzerland

Printed in the United States

Luc E. Weber and Bert van der Zwaan (Editors)
The University at the Crossroads to a sustainable Future
ISBN 978-1-704-29253-3

DEDICATION

To Dr. James DUDERSTADT
Respected scholar, scientist and teacher,
Creative academic leader
Distinguished University President

His colleagues and friends in the organization Committee
dedicate this volume to him, with admiration and gratitude,
for his early enthusiasm, his insights, wisdom, leadership and unfaltering
commitment from 1998 onwards,
all of which has contributed significantly to the development and worldwide
success of the Glion Colloquium.

CONTENTS

PREFACE

The Glion Colloquium held its 12th meeting on 12-15 June 2019 in Glion-above-Montreux, Switzerland. Twenty-two leaders of renowned universities or university organizations participated in the meeting and contributed to the topic proposed by the Programme Committee, *"The contrasting responses of Higher Education worldwide in promoting sustainable development"*. The purpose of the Glion Colloquium 2019 was to deepen and widen the examination of *"The Future of the University in a Polarizing World"*, the subject of the 2017 Colloquium.

The starting point of reference was the observation that the world is presently engaged in a phase of deep and extensive change. The acceleration and broadening of the scientific and technological developments are driving the world into the fourth industrial revolution which has a disruptive impact on industry, services, the labour market and individuals. Combined with globalization, which began in the 1980s after the fall of the Berlin Wall and of the USSR, these two powerful engines of change have contributed to the extremely rapid development of countries which, in the 1970s, were still considered as seriously under-developed. Many countries, particularly in Asia, managed to catch up rapidly on most of their development delay. Some of them, like Singapore, South Korea and China, now compete head to head with leading countries of the West. Globalization and the scientific and technological revolution dramatically increased competition in science and business all over the world. Moreover, the fact that the world is no longer divided between two dominating superpowers, as was the case for decades after the Second World War, has led to a power game between nations wanting to dominate different countries or regions of the world.

These recent developments are in many respects quite favourable for all of those — countries and individuals — directly benefiting from them. But they are not without negative consequences which are now becoming the source of reactions of discontent and of opposition. Modern societies are suffering from rapidly increasing inequality between those who "have" and those who "have not" in economic terms, as well as in terms of individual autonomy. The feeling of a large — and probably growing — minority is that the globalized nature of the economy is harming them, that they have nothing to say or that their votes have no influence whatsoever. They are therefore losing trust in an open and liberal world, in their political system as well as in science and scientists. These feelings are feeding a new rise of nationalist and populist movements, and the development of street movements of opposition, best pictured recently by the *gilets jaunes* in France.

Last but not least, the deterioration of the climate characterized by a rapid increase of the earth's average temperature and more broadly by the increasing negative impact of the world population on planet earth is becoming so visible that a growing number of people feel that that they must do something about it. For those who are conscious that the world population has increased by a factor of 5 since 1900 and that the economic activity of this population measured with the gross national product per head grew by a factor of 5-7 during the same period, it is not surprising if the world population is negatively impacting on the earth's environment and climate.

We could add, not to be complete but because the economic situation worldwide is crucial for the general world prosperity, that the 2007-08 financial crisis condemned the leading Central Banks to inaugurate a totally new policy of cheap money and that they have not been able to return to more conventional policies in 12 years. The consequences are serious as there is presently no agreement on how a potential new recession should be fought.

In conclusion, although it is obvious that scientific and economic developments are extremely valuable for the majority of the world population, the real challenge today is to ensure that these developments remain sustainable in three dimensions: the climate and the environment, society including politics, and economics. If this triple sustainability cannot be secured, the world is certainly heading towards deep new crises.

Universities are obviously crucial actors in the context of these developments. They directly and strongly feed most of the changes that we are observing. Most of the basic discoveries at the origin of the technological changes have been made in their research labs or in independent research labs staffed by researchers trained in universities. They are also training today at least 35% of the specific age class in most developing countries. They are increasingly challenged to keep their quasi-monopoly in higher education and research as they have exercised for most of the last eight centuries. They work very hard to remain at the frontier of knowledge and to meet the growing competition from big corporations, particularly in the computer industry, telecommunication, and life sciences, which enjoy almost unlimited budgets to buy equipment and recruit the best researchers. Moreover, in addition to the continuous increase of traditional students, they have to accommodate returning students and a horde of lifelong learners who come back to university as the knowledge accumulated in their discipline becomes more quickly obsolete.

For all these reasons, universities are by far the best-placed institutions to help governments and societies to solve the problems they are encountering in this period of rapid changes. They are — or at least should be — independent, and the knowledge they develop and transmit is the fruit of a verifiable scientific process open to constructive critics worldwide. As autonomous institutions, independent of government or religious movements, their community is free to choose to study any topic it considers important and relevant. And, in this respect, universities should not only be **responsive** to the changing needs of the population in terms of education and research, but they should also be **responsible** institutions by putting their educational and research potential at the service of societies. This means that, more than ever, universities should not only let their curiosity guide them in choosing their research topics but they should pay increasing attention to the problems which render many of today's developments unsustainable.

In our present time of disarray, governments, business and societies could greatly benefit from relying more on universities in order to help find the best solution, making sure that advances in knowledge could better serve a more sustainable world. At the same time, universities have to improve the way they contribute to solving societal problems without losing their independence and integrity.

Finding a good solution to this double question will be crucial for societies and for universities.

Inspired by the complex and challenging situation described above, the Programme Committee of the 12th Glion Colloquium invited the participants to write a contribution focused on one or more aspects of the chosen theme, and to present and discuss it in one session of the June 2019 Colloquium.

The revised papers published in *"The University at the Crossroads to a Sustainable Future"* provide a striking kaleidoscope of views on the rapid change and growing challenges in the university sector – and their consequences for the purpose and responsibilities of Higher Education and Research. Although most chapters cover different aspects of the general theme, we have structured the book in three main parts, The Global, The Local and The Future. The first one is focused on the changing international context of Higher Education and Research, and the flow of talents; the second one brings together contributions showing that Higher Education should think global, but act local; and the third and last one develops the key role of Higher Education institutions in a sustainable future.

This volume is brilliantly introduced by the keynote address of Michael Møller, Director-General of the United Nations in Geneva and Under Secretary of the United Nations, and concludes with an essay by the editors and the President of the organizing Association on the increased role and responsibilities of Universities and in particular Research Universities to secure a sustainable future.

The XII Glion Colloquium was arranged under the auspices of the University of Geneva and was made possible thanks to generous support from the Swiss State Secretariat for Education, Research and Innovation, the Swiss federal Institutes of Technology of Lausanne (EPFL) and Zurich (EPFZ), the University of Zurich and Nestle, to all of whom we are most grateful. We also wish to thank those who contributed to the colloquium and to the production of this book, in particular Natacha Durand, head of admissions at the University of Geneva, who brought to the meeting her experience of supporting six previous Colloquiums, Dr Gerlinde Kristahn, research fellow, who was the linchpin of the organization, and, last but not least, Edmund Doogue in Perth, West Australia, who provided rigorous editorial assistance. Without these most competent people and generous institutions, the XII Glion Colloquium could not have taken place.

Prof. Bert Van der Zwann
Rector Emeritus University
of Utrecht

Prof. Luc Weber
Rector Emeritus University
of Geneva

CONTRIBUTORS AND PARTICIPANTS

Stephane BERTHET (Co-author of Yves Flückiger's contribution)

Stephane Berthet earned a doctorate in Astrophysics and Astronomy at the University of Geneva in 1991 and then for 11 years contributed to shaping Switzerland's research policies in various fields. Through the State Secretariat for Education, Research and Innovation, he was responsible for representing Switzerland at numerous international research institutions such as the European Space Agency (ESA), the European Southern Observatory (ESO) and Euratom. He returned to UNIGE in 2003 as Secretary General, a position he held through October 2018 and was named Vice-Rector from November 2018.

Leszek BORYSIEWICZ

Leszek Borysiewicz was appointed Chair of Cancer Research UK in November 2016. He became Vice-Chancellor at the University of Cambridge in 2010. Other roles have included Chief Executive of the Medical Research Council and Deputy Rector of Imperial College London. He is a founding Fellow of the Academy of Medical Sciences. Work in vaccines included Europe's first trial of a vaccine for human papillomavirus to treat cervical cancer. He was knighted in 2001 for his pioneering work in vaccines.

Shiyi CHEN

Dr Shiyi Chen became the second president of SUSTech in 2015. Previously he served in roles of Vice President for Research, Dean of the Graduate School and the founding dean of the College of Engineering at Peking University, the Department Chair of Mechanical Engineering at Johns Hopkins University and Deputy Director of the Center for Nonlinear Studies at Los Alamos National Laboratory. Dr Chen is an elected member of Chinese Academy of Sciences and the Third World Academy of Sciences.

Anna DÄPPEN (co-author of Michael Hengartner's contribution)

Anna Däppen is a member of staff and academic associate at the General Secretariat of the University of Zurich (UZH). She took up her present position in May 2016. She graduated from the University of Berne in 2016 with a Master of Arts in Ancient Cultures and Constructions of Antiquity and Prehistoric Archaeology. She holds a BA of Arts from UZH in Classics, Prehistoric Archaeology and French Literature. Between 2012 and 2014, Anna Däppen worked part-time at the Department of Archaeology of the Canton of Zurich and at the Numismatic Collection (Münzkabinett) in Winterthur.

Nicholas DIRKS

Nicholas Dirks is a professor of history and of anthropology at the University of California, Berkeley, where he served as the 10th chancellor until mid-2017. An internationally renowned historian and anthropologist specializing in the study of South Asia, he is a leader in higher education and well known for his thought leadership in areas ranging from the future of the university to the strategic reconceptualization of educational reform on a global scale. Before coming to Berkeley, Dirks was the executive vice president for the arts and sciences and dean of the faculty at Columbia University, and had also taught at the University of Michigan and Caltech. Dirks is the author or editor of seven major books on the history and anthropology of South Asia and the British empire, as well as on a range of themes from social theory to globalization. In February 2018, Dirks was named chancellor and vice-chairman of Whittle School & Studios, a global network of independent schools to be established in China, the United States, India and Europe.

Gérard ESCHER (co-author of Martin Vetterli's contribution)

Gérard Escher obtained his diploma in Biology at the University of Geneva and his PhD (Neuroscience, 1987) at the University of Lausanne, where he led a research group working on synapse formation, after a postdoctoral fellowship at Stanford University. For ten years he worked as Scientific Advisor and Assistant Director at the Swiss State Secretariat for Education and Research. Since 2008 he has served as a senior advisor to EPFL Presidents Patrick Aebischer and Martin Vetterli.

Yves FLÜCKIGER

Since 1992, Yves Flückiger has been a full professor at the Department of Economics of the University of Geneva. Vice-rector of the university for eight years, in July 2015 he was appointed Rector. His research interests focus on the economy of work and of education, particularly on the analysis of unemployment, migration policies, income discrimination and working conditions in different secors. He has authored many books and over 120 articles in international scientific journals.

Alice GAST

Professor Alice P. Gast is President of Imperial College London. Prior to her appointment at Imperial in September 2014, she was the President of Lehigh University (2006—2014) and the Vice-President for Research and Associate Provost and Robert T. Haslam Professor of Chemical Engineering at Massachusetts Institute of Technology (2001—2006). An expert in surface and interfacial phenomena and the behaviour of complex fluids, Professor Gast was a faculty member at Stanford University (1985—2001), being promoted to full professor in 1995. She was affiliated with the Stanford Synchrotron Radiation Laboratory.

Meric GERTLER

Professor Meric S. Gertler is President of the University of Toronto and one of the world's foremost authorities on cities, innovation and economic change. He has advised governments in Canada, the United States and Europe, as well as international agencies such as the OECD and EU. He has authored or edited nine books, and has held visiting appointments at Oxford, University College London, UCLA and the University of Oslo. Among his many accolades, he is a Fellow of the Royal Society of Canada and the Academy of Social Sciences (UK), a Corresponding Fellow of the British Academy and a Member of the Order of Canada.

Michael HENGARTNER

Professor Michael O. Hengartner is currently President of the University of Zurich (UZH) and President of swissuniversities. He studied biochemistry at the Université Laval in Québec City, Canada, and earned his PhD at the Massachusetts Institute of Technology, and led research groups at the Cold Spring Harbor Laboratory in the US and IMLS in Zurich. Professor Hengartner is internationally renowned for his groundbreaking research on the molecular basis of apoptosis.

Timothy KILLEEN

Timothy L. Killeen has been president of the University of Illinois System since 2015. As a space physicist, he earlier served as vice chancellor for research and president of the Research Foundation of the State University of New York, assistant director for geosciences at the National Science Foundation and as a faculty member and administrator at the University of Michigan. He is a member of the National Academy of Engineering and earned his bachelor's degree and PhD at University College London.

Sabine KUNST

Prof. Dr-Ing. Dr Sabine Kunst has been the President of Humboldt-Universität zu Berlin (HU) since May 2016. Previously, she was Minister for Science, Research and Culture in Brandenburg. Before being appointed Minister, she was President of the University of Potsdam (from January 2007 until February 2011). From 2010 until 2011, Sabine Kunst was the first female President of the German Academic Exchange Service (DAAD).

David LEEBRON

David W. Leebron, JD, has served as Rice University's seventh president since 2004, a period of growth and transformation for the university. Prior to that, he was the dean of the Columbia University of Law. Leebron has written in the areas of international trade and investment, torts, privacy, corporate law and human rights. He is the recipient of an honorary degree from Nankai University in Tianjin, China, and has been awarded the title Commandeur de l'Ordre National du mérite by the French government. He is a graduate of Harvard College and Harvard Law School.

Joël MESOT

Joël Mesot was appointed President of ETH Zurich in 2019. Previously, he served for more than 10 years as the Director of the Paul Scherrer Institute (PSI). Dr Mesot studied physics at ETH Zurich and earned his doctoral degree in the field of solid state physics using neutron scattering, both at ETH and the Institute Laue-Langevin (France). Following several years of research in the US, he returned to Switzerland in 1999 to head the ETH and PSI's joint laboratory for neutron scattering. Dr Mesot has been a full professor in physics both at ETH Zurich and EPF Lausanne.

Pratap Bhanu MEHTA

Pratap Bhanu Mehta is Vice-Chancellor, Ashoka University. He has previously been President, Centre for Policy Research. A political scientist, he has taught previously at Harvard and JNU. He was Member Convenor of the Prime Minister of India's National Knowledge Commission in 2005. He has published widely in areas of political theory, constitutional law and Indian politics. Mehta holds a BA (Hons) from Oxford University and a PhD in Politics from Princeton.

Michael MØLLER

Michael Møller has served as the Under-Secretary-General of the United Nations and the 12th Director-General of the United Nations Office in Geneva (UNOG). He was also the Secretary-General of the Conference on Disarmament and the United Nations Secretary-General's Personal

Representative to the Conference. Michael Møller has over 40 years of experience as an international civil servant in the United Nations System, serving in different roles in New York, Iran, Mexico, Haiti and Geneva. Prior to his tenure as Director-General, he was the Executive Director of the Kofi Annan Foundation from 2008 to 2011. In 2019 the Kofi Annan Foundation appointed Michael Møller to its Board.

Andrea SCHENKER-WICKI

Andrea Schenker-Wicki studied Food Engineering and Business Administration and holds a PhD in Operations Research and Information Technology. She headed the section for higher education at the Federal Office of Education and Science and became Professor of Business Administration at the University of Zurich in 2001. From 2012 to 2014, she was Vice President for Law and Economics at the University of Zurich. In 2015, she became President of the University of Basel. Andrea Schenker-Wicki was also President of the Scientific Advisory Board of the Swiss Center of Accreditation and Quality Assurance in Higher Education and a member of the Austria Science Board and the Swiss Science Council.

Atsushi SEIKE

Atsushi Seike is President, Promotion and Mutual Aid Corporation for Private Schools of Japan/Executive Advisor for Academic Affairs (President Emeritus), Keio University. A labour economist, Professor Seike has been a Member of the ILO Global Commission on the Future of Work, President of the Japan Society of Human Resource Management, Chairman of the National Council on Social Security System Reform and President of the Japan Association of Private Universities and Colleges. His current positions in government include Chairman of the Council for the Promotion of Social Security System Reform and Honorary President of the Economic and Social Research Institute.

Michael SPENCE

Dr Michael Spence has led the University of Sydney as Vice-Chancellor and Principal since 2008 and is the University's 25th Vice-Chancellor. An alumnus of the University of Sydney, Dr Spence has a BA with first-class honours in English, Italian and law. His other languages include Chinese and Korean. He holds a Doctor of Philosophy from the University of Oxford and headed Oxford's law faculty and its social sciences division. During his time as Vice-Chancellor, the University of Sydney has forged a distinctive, new strategy focused on the transformation of undergraduate education, promoting interdisciplinary research, and strengthening the culture around its core values. In 2017, he was awarded a Companion of the Order of Australia in the Australia Day Honours for service to leadership of the tertiary education sector, to the advancement of equitable access to educational opportunities, to developing programs focused on multidisciplinary research and to the Anglican Church of Australia.

Subra SURESH

Subra Suresh is President and Distinguished University Professor at Nanyang Technological University, Singapore. A former Director of the US National Science Foundation, he now serves as an independent Director of the Board of HP Inc. (HPQ) and Singapore Exchange (SGX). He has been elected a member of all three branches of the US National Academies — Engineering, Sciences and Medicine — and a foreign member of science academies in China, France, Germany, India and Spain. He has been awarded 18 honorary doctorates from institutions around the world.

Eng Chye TAN

Professor Tan Eng Chye was appointed President of the National University of Singapore (NUS) on 1 January 2018. A pioneer architect of the current academic system in NUS, Professor Tan has seeded many initiatives such as the Special Programme in Science, University Scholars Programme, University Town Residential College Programme, Grade-free Year and Technology-enhanced Education.

Martin VETTERLI

Researcher, teacher and expert of the Swiss education and research landscape, Martin Vetterli was appointed president of the École polytechnique fédérale de Lausanne (EPFL) in 2017. He is also a world renowned expert in the areas of electrical engineering, computer sciences and applied mathematics, and a full professor at the audiovisual communications laboratory at EPFL. From 2013 to 2016, he was President of the National Research Council of the Swiss National Science Foundation.

Luc E. WEBER

An economist and professor of public economics at the University of Geneva, Luc Weber served for more than 30 years in Higher Education and Research in Switzerland, Europe and the wider world. Vice-Rector and Rector of his University and President of the Swiss Rectors' Conference, he then served numerous international university organizations, governmental and non-governmental, European and worldwide: President of the Steering Committee for Higher Education and Research of the Council of Europe, Vice-President of the International Association of Universities and founding Board Member of the European University Association. His excellent knowledge of the sector inspired him to create and conduct, from 1998 onwards, the Glion Colloquium.

Jaeho YEOM

Professor Jaeho Yeom served as 19th president of Korea University (2015—2019). He earned his PhD in political science at Stanford University for his research on Japanese industrial policy for high technology. He has taught public administration at Korea University since 1990. Professor Yeom has also performed research as foreign visiting professor at Tsukuba University in Japan, visiting professor at Griffith University in Australia and Beijing University in China, adjunct professor at Renmin University in China and Chevening Fellow at CENTRIM at the University of Brighton in the UK.

Tomáš ZIMA

Prof. Tomáš Zima, MD, DSc. Dr.h.c. graduated from the First Faculty of Medicine of Charles University and is a professor of medical chemistry and biochemistry; he specializes in clinical biochemistry, internal medicine and nephrology. Since 2014 he has been Rector of Charles University and President of the Czech Rectors' Conference, as well as being a member of UNICA Steering committee and of EUROPAEUM's Board of Trustees. He is a member of the European Commission's Scientific Panel for Health (SPH).

Bert van der ZWAAN

Bert van der Zwaan is emeritus professor of Biogeology at Utrecht University in the Netherlands and was Rector Magnificus (Vice Chancellor) of Utrecht University in 2010-2018. He was a member of the board of directors and president of the European League of Research Universities (LERU, 2013-2018) and is author of the book, *Higher Education in 2040* (2017). Since retiring, he has written about higher education. He is nationally and internationally active in supervisory and advisory boards, advisor to many universities in strategy planning and independent chair of the energy transition debate in the region of Utrecht.

KEYNOTE ADDRESS

Remarks

Michael Møller

We are meeting on the eve of your colloquium, which is why — when I was kindly asked to kick-start our discussion this evening — I thought it would be constructive to take a step back and begin with more of a bird's-eye view. Specifically, I would like to start by talking about the state of the world as seen from the vantage point of the United Nations; to trace the evolution of how we arrived at the present moment; what it teaches us about what we need to do next; and, most importantly, to connect it all with the role, responsibility and promise of universities.

THE STATE OF THE WORLD

Start with the state of the world. I am often invited to speak to young students across the world, and I am always intrigued by a paradox they are facing.

On the one hand, they are seeing a world in deep crisis, a world that — ecologically, economically, politically — seems to be teetering on the brink of collapse.

They see a climate crisis wreaking havoc. Armed conflicts threatening millions and refugee flows at record levels. Rampant inequality both between and within countries. Escalating disputes over trade, sky-high debt, threats to the rule of law, attacks on the media and civil society. These ills affect people everywhere and they are all connected: climate disasters entrench poverty; poverty breeds conflict; conflict triggers refugee flows, and so on and so forth. Together, these threats are deeply corrosive. They generate anxiety and they breed mistrust. They polarize societies – politically and socially. And so we see many people turning their back on the "system". And to be sure, not without cause:

- Can you blame people for questioning the legitimacy of an order in which 26 men own as much as the almost 4 billion people who make up the poorest half of the world's population?
- And can you really expect today's students to be optimistic about the future, if their generation faces — for the first time in a long time — the very real risk of earning and owning less than their parents?

Against these questions, explanations often sound like excuses — and it is not difficult to understand why faith in political and business leadership is waning; why trust in national governments and international organizations is eroding; and why populism is gaining traction.

But I mentioned a paradox a moment ago, and it is essentially this:

Against the doom and gloom of our time, there is a powerful counter-point. By virtually every measure of well-being, humanity is better today than at any other time in its history.

It's a fact. Living standards, life expectancy, literacy rates and education levels have never been higher across the world. Child mortality, the risk of dying from disease, from war or famine, has never been lower. Today, for the first time in history, infectious diseases kill fewer people than old age; famine kills fewer people than obesity; and violence kills fewer people than accidents.

All of this and much more happened over the course of just a few decades. And all that progress is real. It has been broad, and it has been deep, and it all happened in what – by the standards of human history – was nothing more than the blink of an eye. And now an entire generation — the generation of your students — has grown up in a world that by most measures and in most places has become steadily healthier and wealthier and less violent and more tolerant during the course of their lifetimes.

None of this is to suggest that things are just fine. They clearly are not. Rather, these data points highlight an intriguing contradiction, namely that we seem to be both living in the best of times and in a time of existential crisis.

How do you reconcile it?

EMERGENCE OF MULTILATERISM

I think the answer has a lot to do with the challenges faced by the organization that I work for, the United Nations. And more generally: with the fate of the multilateral system and the very idea of international cooperation.

Let's unpack it by going back in time. By going back, in fact, exactly 100 years. The First World War marked a watershed in many ways, and one of them was the bankruptcy of the old idea that balance-of-power politics could ever be a guarantor of peace.

Clearly, an alternative international order was needed and in this vacuum emerged the idea of multilateralism, finding expression in the League of Nations in Geneva. To be sure, the inability of the League to prevent a second world war has long made it a byword for failure, a graveyard of hopes.

Today, however, that simplistic, unfair view is giving way to the recognition that the League — despite its constraints and contradictions — nurtured the nucleus of a system that has since proven to be remarkably successful. For when the United Nations was created out of the remnants of the League in 1945, the multilateral order finally caught on.

The audacity of the ideas that underpins the multilateral architecture remains astounding: to replace violence with the rule of law as the basis for global governance; to give each state — whether rich or poor, large or small — one vote; and finally, to declare human rights unconditional and universal.

Of course, there were many places in which reality made a mockery of the ideal, where tyrants still ruled; where colonial regimes refused to give way to the forces of freedom. But they soon found themselves on the defensive.

And of course, the Cold War, and with it the terrible nuclear threat, cast a long shadow. But not only did we avoid open confrontation between the superpowers — and with it a third world war — war itself came to be considered "illegal", an idea that would have seemed simply absurd to earlier generations.

And with these political changes came sweeping economic changes — leading to the incredible gains in global wealth, in life expectancy and opportunity, that I mentioned earlier.

It's no accident of history that the progress we achieved since 1945 coincided with the establishment of the multilateral order with the United Nations at its heart. There is a direct connection here. You can see the connection in measures small and large. Let me just mention three out of thousands of examples:

1. You can see it in conflicts prevented or defused across the world by the quiet workings of UN mediation — in places as different as El Salvador, Sierra Leone or Nepal.

2. You can see it in deadly diseases eliminated by actions led by the World Health Organisation — like the vaccination programs that eliminated smallpox.

3. You can even see it in the dialling codes you use to call colleagues and family abroad — a system developed by the International Telecommunication Union down the lake in Geneva.

All of the above is multilateralism in practice.

And yet, for all the peace and prosperity underwritten by the international structures put in place since 1945, today, we once again find ourselves engulfed in crisis. So what happened?

A NAÏVE BELIEF

Sometime over the past decades, a complacency set in — a naïve belief as it turned out — that things would just invariably get better; that, despite some backsliding here and there, forward movement was inexorable and large-scale conflict a thing of the past. It was through this lens that many just assumed technological progress and globalization would produce benefits that, ultimately, would reach all.

This complacency bred inaction, and the twin forces of globalization and technological disruption — left unchecked — ultimately triggered the global backlash we are confronting today.

And so today, we hear troubling echoes of the past.

Some of these "echoes" I have alluded to already — from eroding trust in the democratic order to the outrage of rampant inequality. But the one I want to explore further has to do with the breakdown of global cooperation, with the return of international politics as zero-sum competition.

Today, we no longer live in a bipolar or unipolar world; we are increasingly in a multipolar world. And we are in a chaotic transition phase. The relationship between the three most important powers — Russia, the United States and China — has rarely been as dysfunctional as it is today.

And, related to that, medium-sized powers are increasingly acting autonomously from the big powers. It's impossible to look at Syria, for example, and not recognize the role of Turkey, Iran and Saudi Arabia. And the same is true for other conflicts around the world.

So power relations are becoming unclear; with the fragmentation of actions; with impunity and unpredictability prevailing; and with national and isolationist agendas superseding mutual trust and international cooperation.

The point here is that we have been there before — and that should worry us. Because multipolarity without strong and accepted multilateral instruments — just as we saw in Europe in the wake of the First World War — might be a factor of some equilibrium, but it is certainly not a factor of peace.

It's inherently unstable, volatile, and dangerous. So that is another echo of the past we hear today. Yet to say that the world is poised on the brink of another 1914, as some suggest, is too simple.

International relations work differently today, and so does politics.

AGE OF ENTANGLEMENT

One obvious difference is the diffusion of power. Power that used to be firmly in the hands of the state has metamorphosed into something much more diffuse: whether it's non-state actors challenging the state's monopoly of violence; or whether it's private corporations evading effective regulation by any one state — power in international relations today is altogether a more complex, messy affair.

One way to think about this change is as a contrast between hierarchy and order versus networks and entropy.

Whereas in the past, international relations were centralized — with core and periphery, with top-down commands and control — today, we live in an "age of entanglement".

Global politics has been reconfigured. The traditional "chessboard" of inter-state diplomacy may still exist, but it is joined by a new web of networks made up of governments, companies, NGOs, terrorist groups, philanthropists — university rectors — and countless others — all wielding influence and cooperating or clashing at various points in time.

In response to all of this, multilateralism is changing, too. By necessity, it has to become more integrated, more networked, more inclusive — and the upshot is that everyone in this room today forms part of the networks that will define the way multilateral global governance will evolve. And these intricate connections are mirrored by the major existential challenges we face, which, as I said at the outset, are more and more interlinked; are more and more interfering with each other.

Let's take stock: We're facing a crisis of trust, challenges threaten to overwhelm us just as interests fragment, power is diffuse, and the only constant is disruptive change.

Where do we go from here? How do you react? Those are — in the broadest sense — the questions that have brought us together today. And the answer has everything to with the 2030 Agenda for Sustainable Development.

AN AMBITIOUS AGENDA

It is, quite simply, the most ambitious development agenda in human history, agreed by all 193 Member States of the United Nations.

We now have a detailed roadmap of what needs to be done.

It is firmly built on the following three bedrock principles: that the 17 goals are indivisible: you cannot deal with one or two of them without keeping the others in mind; that they leave no one behind and that everyone — the private and the public sector, academia, civil society, the rich countries and the developing countries, every individual – is responsible for achieving them.

The 2030 Agenda gives us the substance and the practical philosophy for a multilateralism fit for the 21st century — networked, collaborative, and inclusive.

It is our global blueprint for creating an adaptive — and agile — coalition that can respond at speed and at scale, something that neither national governments, nor individual companies, nor anyone else can ever hope to achieve in isolation.

Given that the stakes could not be higher, everyone needs to take a hard look at themselves and see whether they are part of the solution, or part of the problem. This clearly is no time for bystanders.

What does it mean for universities?

EDUCATION A CORE ELEMENT

There are many ways to approach this, as there are many ways in which the contribution of universities is absolutely critical.

First, your role as providers of education. Education is the currency of the Information Age, no longer just a pathway to opportunity and success but a prerequisite. At the UN, we are spending a lot of effort on leveraging our actions to have the greatest long-term impact. That means not just chasing the latest headline-grabbing emergency, but tackling the root of the problem; addressing the cause, not just the symptom; it means focusing on prevention.

Indeed, the 2030 Agenda is above all a prevention agenda. And education is an integral, core element. That is why education is both a stand-alone goal (Goal 4) and linked either explicitly or implicitly to virtually all of the other SDGs.

Achieving equitable economic growth; reducing social and gender inequalities; empowering marginalized groups; driving innovation; promoting tolerance; enabling a life of dignity — any one of those begins and ends with one thing: education.

The question of course is — and this strikes me as particularly relevant for institutions of higher education — with all the disruption that already happened, and the disruption that is on the horizon; with so much that has changed: has education changed enough? Are you preparing your students in the best possible way for the world that's around the corner?

I do not of course have clear answers to those questions, but I know that they go beyond the world of academia.

They are questions of content: what are the skills and the knowledge that will be critical going forward? Is it really just the sciences, is it nanotechnology or bioengineering? What if it is the humanities which actually teach you the adaptive, transversal skills that best position you to manage the disruption ahead?

A liberal education — as defined by Cardinal Newman in the 19th century — is a "broad exposure to the outlines of knowledge" for its own sake; it teaches you how to learn.

Looking at the ways in which technology and globalization are transforming our world, five years from now, your graduates may very well be working at a company that hasn't been founded yet. In 10 years, they may work in an industry that doesn't exist today. So that's why curiosity and interdisciplinarity are so important: an ability to connect the dots across disciplines; to think holistically; to break down silos; an interest in other cultures, an appreciation for different viewpoints. The very principles the SDGs were built on.

Which is why we need to get much better at devising and implementing curricula that promote an integrated, transversal and multidisciplinary approach to education.

If we used to take the past as a guide for the present, today, we increasingly need to use the future. What will matter most will be to "learn how to learn", much more than to learn lots of things. And it is clear that life-long learning will be the centre of education and training systems for vast segments of society.

And that means the questions you are posing yourselves over the coming days are also questions of accessibility:

Millions of jobs will disappear; millions of jobs will be created — but the vast majority of them will require some form of higher education. Universities — by becoming more open, more affordable, more inclusive, more flexible — will play a crucial role in the success or failure of our ability to manage the years and decades ahead.

Then there is the question in how far the value of an education should be measured against the yardsticks of ethics. Do universities have a responsibility to instil an ethical compass in students?

An early Facebook employee once famously remarked that "the best minds of my generation are thinking about how to make people click ads" — which, however lucrative or intellectually challenging a profession it may be, we can all agree it does not tackle the urgent threats facing humankind.

And everything I said about your role as providers of education is true also for your role as centres of research. It is not just that the research you fund and undertake will determine our ability to combat climate change, to harness the potential of technology for good, to fight diseases, and much more.

Your influence — and by extension, your responsibility — extends even further than that.

Preparing my remarks, I was reminded of the observation of John Maynard Keynes, who once said that: "Madmen in authority, who hear voices in the air, are distilling their frenzy from some academic scribbler of a few years back. Indeed", he went on to say, "the world is ruled by little else."

So whether he was purposefully exaggerating here or not, the fundamental point still stands:

The complacency in recent decades I mentioned earlier — was it not the upshot of a belief in the promise of unregulated free markets that first emerged in academia?

And by the same token, the comprehensive shift towards sustainability — does it have a chance if it is not buttressed by academic thought?

All of which is to say: your role and responsibility in our collective efforts in the face of truly existential challenges are enormous. We have the means and the skills to create a world that is fairer and more peaceful, that is ecologically sustainable, and in which the incredible riches of our world benefit not just the fortunate few, but lift the fate of the many.

But we can only hope for success if every single one of us fully commits and buys into this effort. And we need to get better at acting together, we need to start speaking the same language and work towards the same goals. That is what will make or break our whole endeavour.

Thank you and I look forward to our conversation.

PART I

• • • • • • • • •

The Global

CHAPTER 1

The Geopolitics of Higher Education

Yves Flückiger and Stéphane Berthet

INTRODUCTION

T oday, Higher Education Institutions are not only globalized but also globalizing entities. This evolution has gradually formed over the last three decades, changing profoundly the world landscape of academic institutions, which has become a competitive market.

Academic globalization most often refers to the increasing openness of universities to exchanges, student and researcher mobility, the multiplication of strategic partnerships and the harmonization of curricula and degrees. This globalizing dynamic takes place in very diverse contexts from economic, social and historical points of view. What common characteristics exist today between the major classical research universities in the Top 100 of the Shanghai ranking and a university located in a developing country that has to manage large numbers of students and where research activities are often non-existent? In addition, the gap between those different universities tends to widen since globalization increases the dynamics of inequality and reinforces the logic of competition.

One important factor in university globalization is the access of an ever-increasing proportion of the population to higher education. Over the past three decades, the number of students worldwide has almost doubled every 10 years, from 50 million in 1990 to more than 215 million today, probably reaching 380 million in 2030 (Vetterli & Escher, this volume). Remarkably, the centre of gravity of the student population has shifted. Since 2003, there are more students in so-called emerging and developing countries than in

OECD countries. Most of this changeover was due to China and India, which now have more than 50 million students. China, in particular, has put in place a strategy to encourage its best students to train at the best universities in the world and then offer them very attractive jobs and high salaries to encourage them to return home. Therefore, it is no coincidence that it is precisely China that has set up the first international ranking to identify the universities to which its students should be sent.

Actually, university rankings play a crucial role and are an important indicator of the power of universities to attract the best students and the most productive researchers. They are also indicators for the economic health of countries, not only because they point out the capacity of nations to invest into higher education, but also because the ratio of public expenditures on education to the GDP is a factor of economic success. As part of what is sometimes called smart power, knowledge, and more precisely higher education, appears to be both an index of influence and a power factor. From this point of view, the international rankings that have abounded over the past 20 years have played a major role in this reconfiguration of the university landscape on a global scale. We will therefore first ask ourselves why these rankings appeared and how they were established.

GOING GLOBAL: WHY?

A number of trends are responsible for driving change across higher education and university-based research over recent decades. There are two broad dimensions: the changing social contract between higher education and society, and the geo-politics of knowledge in a globally competitive world. Globalization has partially transformed higher education and the increasing reliance on knowledge for economic competitiveness has obliged the state to remain involved in higher education, even as it purports to withdraw from other spheres through privatization. While science has always operated in a competitive environment, the emergence and increasingly persuasive role of global rankings has made the tension between national and global ever more apparent.

The rise of neo-liberalism and corresponding adoption of principles of new public management are credited with changing the relationship between higher education and the state, and between the academy and the state. This led to more autonomy focused on performance-based funding or performance agreements.

The birth of international rankings was marked by the advent of Shanghai Jiao Tong Academic Ranking of World Universities in 2003. But their true origins lie in the growing tension between the role of knowledge for global competitiveness and, correspondingly, the national social contract with higher education and science. International rankings are a product of an increasingly globalized economy and an internationalized higher education landscape, which has become a competitive market. These rankings affirm that in a globalized world, with heightened levels of capital and talent mobility, national pre-eminence is no longer sufficient. Despite considerable scrutiny and criticism over the years, rankings have persisted in informing and influencing educational policy, institutional funding, academic behaviour and stakeholder opinion.

The emergence of a global knowledge society poses new challenges for universities, which are places of creation, innovation and knowledge transmission. The United States and Japan, as well as India and China, have understood this and have massively increased support for university scientific research in recent years. Switzerland, whose position is still enviable, must meet these challenges in a context that could become more difficult if it were to isolate itself from the European Union. More than ever, its socio-economic development depends largely on its ability to train the many young people sorely needed by our country, lacking any other natural wealth, to ensure lifelong learning and to foster an evolution of society that enables it to respond to the changes it must face. The University, through its ability to develop world-class research centres, is an absolutely necessary instrument for this socio-economic development.

But to maintain its position in an increasingly competitive international environment, universities must cultivate their excellence. Not for themselves, but to make their essential contribution to the region in the area where they are located. Even if the very concept of excellence, and the criteria for measuring it, are often criticized, they nevertheless make it easy to compare the different universities around the world and thus constitute, for young people, an often important element in their choice of place of study. They also send a strong signal to employers who make a first selection based on their candidates' applications; and undeniably influence the ability of academic institutions to raise donor funds.

Typically, the Shanghai ranking exclusively measures the quality of a university through cutting-edge research, whether fundamental or applied. Although it is regrettable that there are no criteria related to teaching or other aspects of university excellence, this ranking has the advantage of being based on objective data collected by the rating agency and not by the universities themselves. However, the main criticism of this ranking remains that it tends to ignore disciplines that do not award Nobel prizes or that do not have access to scientific journals such as *Nature* or *Science*.

More than the precise rank obtained in rankings, universities must maintain their position among the 200 best universities in the world. It is at this level that the competition to attract talent from all over the world is making it possible to nurture the training and research of an academic institution. Otherwise, a university can no longer play its role as a driving force for economic and social innovation, which is a pillar of the competitiveness of its region. This need is clear from the analysis presented in Figure 1, which shows that the number of citations per published article in the life sciences field is a decreasing function of the rank obtained by a university (Van Raan, 2005). Interestingly, Figure 1 shows that beyond the 200th rank, the number of citations drops sharply, demonstrating the loss of impact that these articles have on the development of the life sciences.

Figure 1 – Correlation between impact of top universities
in the life and biomedical sciences (CPP) and ranking position (r)

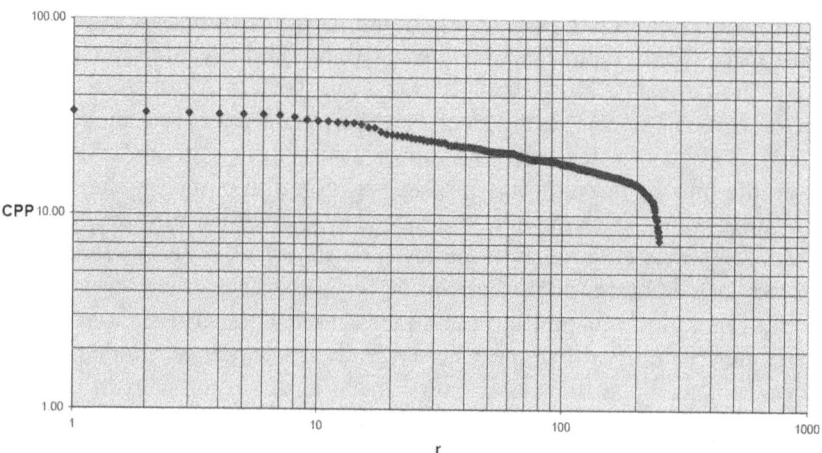

The quality of training is closely linked to the quality of research. This requires attracting the best talents. The excellence of researchers is a necessary condition for academic institutions to remain at the level of the best research centres throughout the world. It is from competition between the elite of researchers that the most spectacular scientific advances and innovations necessary for the economic development are born. It is also through their collaboration and the shared use of advanced, complex and costly infrastructure that science is advancing.

China in particular has adopted a highly geopolitical strategy to bring its best universities to the top of the Shanghai ranking, a strategy that was quite successful considering Figure 2 which shows the ranking by countries according to the number of national universities belonging to the top 100 best institutions. The Chinese student diaspora has grown steadily in recent years. According to the Chinese Ministry of Education, in 2010 1.27 million

Chinese students had gone abroad to study. In comparison, in 1990 only 7,647 had been sent abroad to study (China Education Yearbook Editorial Board, 1991). At the same time, the Chinese government is encouraging these students to return to China and attract many foreign students to train them. This scientific community has an influence on investment decisions in Research and Development, whose weight in relation to GDP has more than tripled over the last three decades, from 0.56% in 1996 to more than 2% in 2015, to reach the objective of 2.5% in 2020, the level reached by the United States.

In this perspective, international rankings play an important role on the prestige and attractiveness of global universities. This is why the European Commission has decided to create its own index, the U-Multirank, whose objective, more or less admitted, is to promote European universities. The spread of Shanghai's ranking throughout the world, both in the media and in the political sphere, has made university rankings a powerful tool that goes far beyond the academic field.

Figure 2 – Evolution of rankings for some countries
(Academic Ranking of World Universities-Shanghai Ranking Consultancy)

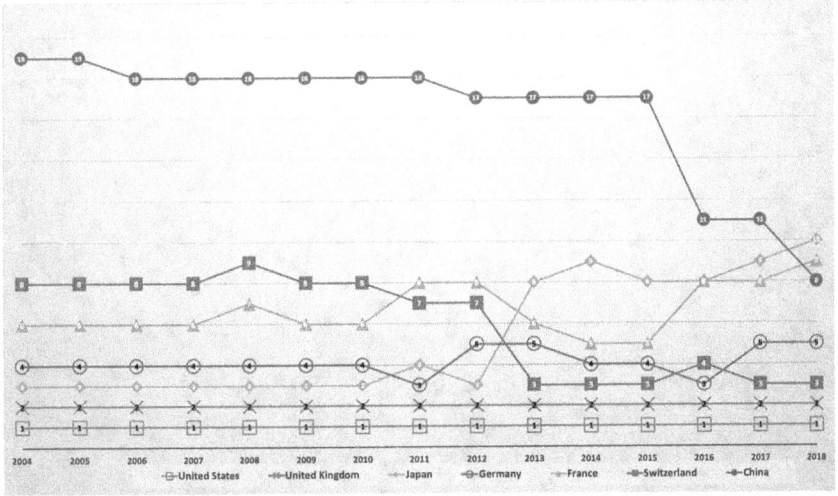

If China's strategy was successful in terms of this policy to bring its best institutions among the top 100 universities, Figure 3 however shows that there is wide disparity within the Chinese academic system. Only 5% of the 62 Chinese universities ranked in the top 500 in the world belong to the top 100, a ratio that stands at 100% in Singapore and 63% in Switzerland. This result shows that the majority of students in the latter two countries benefit from a quality education.

As illustrated by Figure 3, one of the criticisms frequently made of international university rankings is that they contribute to increasing inequalities between mass and research universities but also, and even among universities in developed countries, between academic institutions according to the total amount of their budget or the amount of their budget per registered student. From this point of view, it is interesting to consider the revised Shanghai ranking by weighting the results by the size of the budget, first, and then by the size of the budget per student. Tables 1 and 2 in the appendix highlight the upheavals brought about by these new approaches. They show the more or less efficient use made by universities of the resources allocated to them, putting all institutions on an equal footing regardless of the size of their budget — and it should be noted that the usual numbers 1 and 2 are no longer even in the top 50. (Olivier Berné, CNRS and Université de Toulouse, https://nouvellesdesetoilesblog.wordpress.com/2018/08/17/le-classement-de-toulouse-des-universites/).

Figure 3 – Ratio between the number of universities
in the Top 100 and number of universities in the Top 500
for all countries that have at least one university in the Top 100
(Academic Ranking of World Universities-Shanghai Ranking Consultancy)

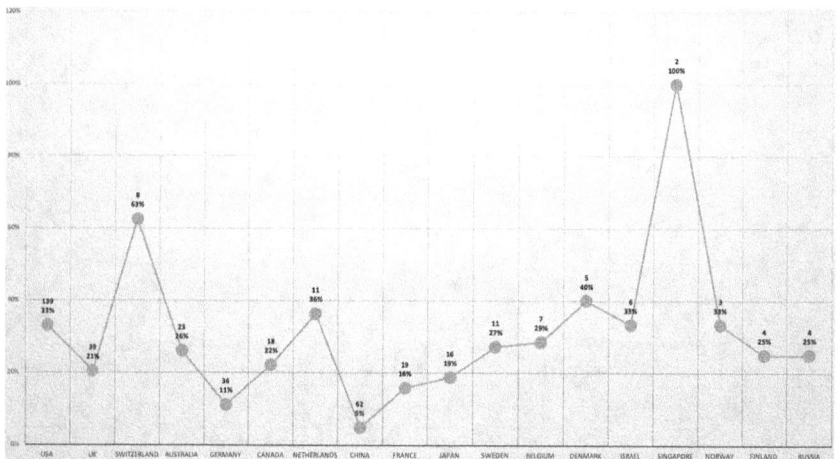

GOING GLOBAL: HOW?

Some facts about the internationalization of universities

Academic mobility (students and faculty) is a tradition that dates back to the creation of universities and it is certainly the most frequently considered example for the internationalization of universities. Nevertheless, since the

1990s other elements have also taken place in this context, such as the internationalization of curricular and the development of university partnerships.

As can be seen in Figure 4, student mobility has intensified with the globalization of the higher education sector. The goal of the Bologna Process was precisely to create a European Higher Education Area, with comparable institutions in terms of standards and quality of higher education qualifications to facilitate academic mobility.

Figure 4 – Growth of international students worldwide 1975-2013

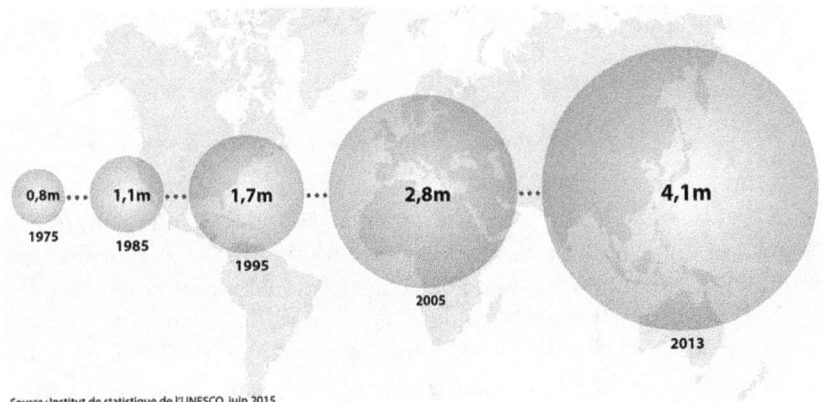

Source : Institut de statistique de l'UNESCO, juin 2015

Indexes and metrics are a valid and necessary starting point for the analysis of globalization and student mobility. However, in the end, it is all about people. When individuals decide to pursue studies in a foreign country, they do so in the hope of being exposed to an experience that will nurture their lives and help them build a better future for themselves and their families. Cultural values are rapidly changing and the younger generations are realizing that international mobility dramatically increases the number of opportunities available for individual advancement.

International mobility is also organized in a competitive mode where universities compete for the best professors and students, which has an impact on the geopolitical map of higher education. Thus, it is not surprising that leaders of technological companies such as Bill Gates consider that the only way to solve the US "critical shortage of scientific talent" is to open up the visa system to special categories of immigrant workers. This competition to attract talents is illustrated by Figure 5, especially at the level of PhD Students for countries like Switzerland lacking highly qualified people to sustain their economic development.

Figure 5 – Percentage of foreign students at different level of the curriculum

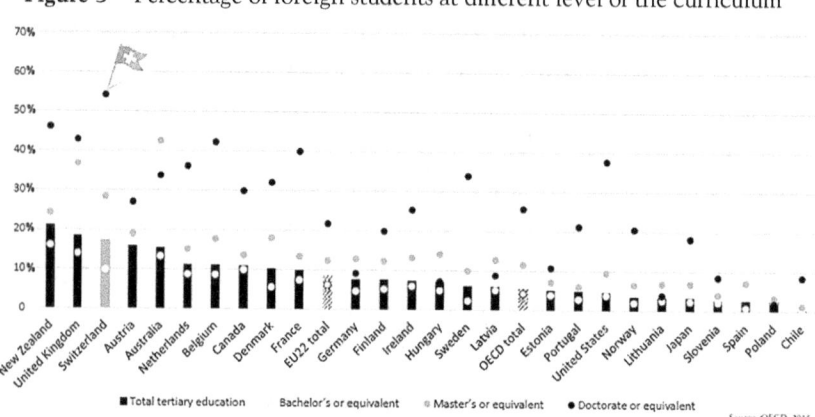

To the above arguments regarding competition for talents, it could be added that higher education is more and more considered an important diplomatic asset contributing to a reduction of friction between countries and cultures. It is not surprising to see new organizational initiatives linking foreign policy with international student mobility and academic cooperation, e.g. the creation of the Bureau of Educational and Cultural Affairs at the US Department of State or locating outposts of Campus France in French embassies around the world.

Interestingly, students from developing countries present a higher willingness to move between national borders than those from developed nations. This "cultural melting pot" poses a challenge for host countries. Although at first glance it may appear the students coming from developing countries are being unilaterally exposed in the cultural waters of industrialized nations, it is also true that the incoming cultures are transforming the receiving countries' behaviours. Well-established institutions attracting an increasing number of international students are already facing a dilemma of balancing their own "traditions" — the ones that took them to the leading position they occupy today — with the need of internalizing the cultural baggage brought in by international students. International mobility may be accompanied by turmoil, but it is a challenge that any country and any university wishing to excel in the higher education arena cannot avoid.

Internationalization has also reformed curricula with the aim of injecting an international element into the content and delivery of programmes. The most prominent (though possibly not the most frequent) form of curricular internationalization is the delivery of a program in a language other than the one of the country where this programme is offered. In the vast majority of all cases in Europe, this language is English. English-medium provision in Europe has seen a strong growth in the last five years, even though it still constitutes only a fraction of all provision in European higher education. What makes this form of education international is, first and foremost, the language of delivery, and — second, and only related — the (usually) international composition of the student body. In addition to this, they are international curricula, which are jointly delivered by two or more higher education institutions in at least two countries. More recently, new forms of internationalization appear consisting in a variety of manifestations, from branch or off-shore campuses to delivery abroad of programs with the help of a (licensed) foreign tertiary institution, and various forms of distance (usually online) education offerings, to name only some. The common feature of all these is a particular form of mobility, in which it is not the student that moves across a country border, but the educational offering.

Universities are also faced with the necessity to build international partnerships and establish mobility pathways which carry both knowledge and social impact, which contribute to social growth as well as institutional growth. It should be noted that there is also growing internationalization in the context of "quality", evidenced, not in the least, by the attention accorded to international rankings. In 2018, the International Association of Universities (IAU) conducted its fifth global survey and it appears that the two most important benefit of internationalization are "enhanced international cooperation and capacity building" and "improved quality of teaching and learning".

The key challenge facing Higher Education Institutions is not only to monitor and track partnerships beyond the agreed memoranda of understanding, but to build and to sustain mobility and internationalization, through the resourcing of intelligent solutions, trend analyses and performance data which can be leveraged into institutional strategy for growth, excellence and impact in an ever-changing world. It is not a surprise, then, that Asian countries — particularly China and India — are the main source of internationally mobile students, while Western countries with solid higher education systems lead the way in terms of inbound flows.

If, for the time being, there has been uncontested dominance of North American and European higher education, together with Japan, it is only a matter of time before this lead starts to diminish. The enormous, long-term growth-potential, combined with a favourable political climate for economic development, makes it inevitable that by the middle of this century higher education in other regions will catch up in every way that matters in their respective economic progresses. European universities, particularly those functioning in systems with generous public support, have a mixed attitude in accepting the new paradigm of global academic competition and advocate the status quo combined with an increase in public funding.

Researchers' mobility

If we look now at researchers' mobility, it can be observed that Europe is high in mobility with much intra-region movement, while Asia has more outbound movement, mostly to Americas, followed by Europe, and then Oceania. The Americas have more inbound movement, from Asia, Europe and same region.

Figure 6 – Mobility in and out of Researchers

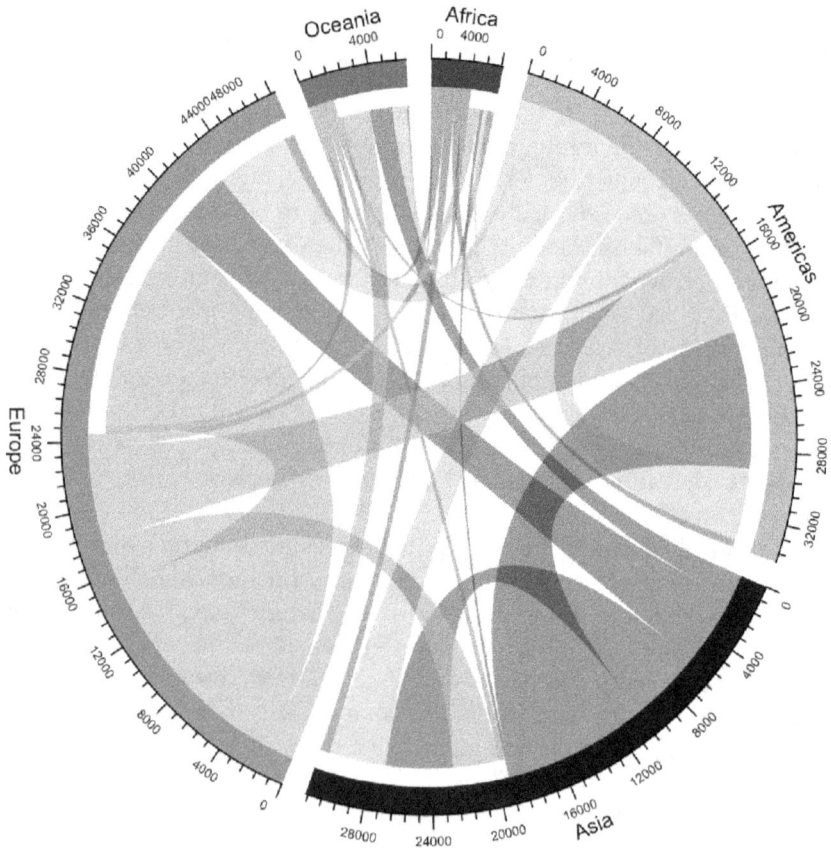

Figures 7 and 8 represent the ratio between researchers migrating out of a given country in comparison with researchers moving in. Without any surprise, beside India, China has the second-highest ratio, losing five times more talents than gaining them. This is fully in line with the Chinese strategy to build a higher education system based on researchers educated abroad. In terms of attracting researchers, Qatar and Saudi Arabia are getting the most influx compared to very little outbound movement. Singapore and HK in Asia, as well as Switzerland, are also attracting 2–3 times more researchers than losing them.

Figure 7 – Highest Outbound/Inbound ratio in research migration

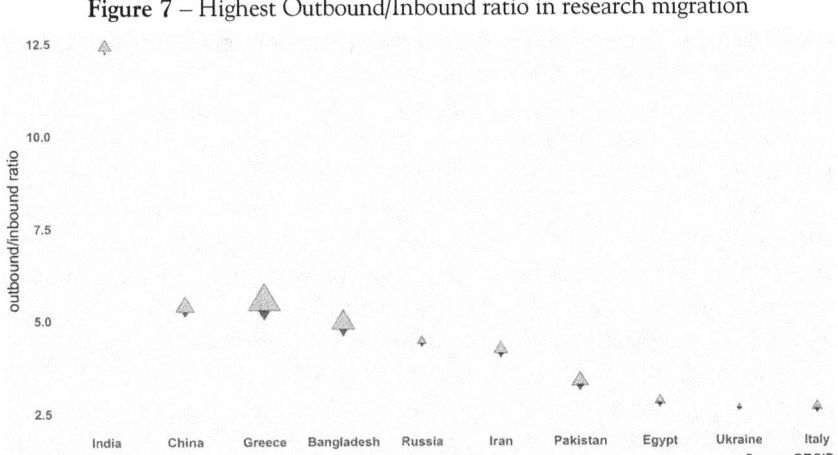

Source: ORCID

Figure 8 – Lowest outbound/inbound ratio in researcher migration

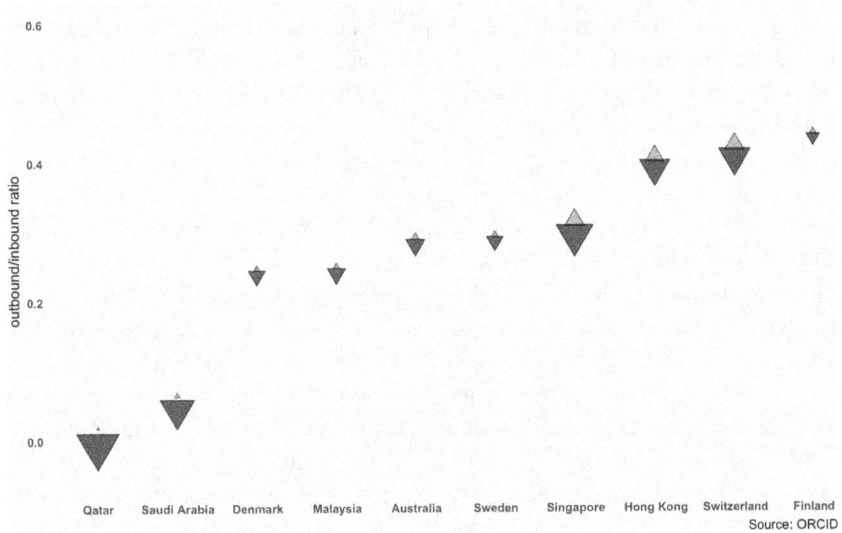

Source: ORCID

THE GEOPOLITICS OF RESEARCH
AND THE EFFICIENCY OF PUBLIC POLICIES

Publications as a key element of rankings

Nowadays, the big funding for research is allocated only to the best projects. Good is not sufficient. For this reason, we can talk of a new quality regime: "moving from good to excellence" in higher education and research policy. This new quality regime combined with searching for excellence also raises one of the most challenging problems related to institutional configuration of European (continental foremost) higher education. Such a regime may question the system-wide validity of the so-called Humboldtian model of the university which has been the dominating conceptual and organisational framework for higher education in Europe for almost two centuries. This model puts "research" at the heart of the traditional university and is naturally linked to teaching, thus assuring a mutually reinforcing mechanism for the free circulation of knowledge between research and teaching. We do not consider that the Humboldtian model is altogether obsolete, but it does only reflect a certain type of higher education institution, which is often referred to as the "research-intensive" university.

Can we afford for all institutions to be "research-intensive"? No less important for our analysis is that research has become a highly globally competitive activity, which requires enormous investments in personnel, infrastructure and equipment. Therefore, when trying to adapt themselves to "the global battle for intelligence", countries are introducing a preferential system for supporting research excellence, recognizing that only through a competitive approach and a steady level of appropriate funding are they going to be better positioned for meeting the future challenges of higher education and research. In this context, publications, as the ultimate output of research, play a crucial role for in terms of geopolitics of higher education.

During the last 20 years, the evolution of the number of papers published by Chinese universities is impressive. In a recent article, Xie and Freeman (2019) measure countries' contribution to the world scientific literature according to the authors' addresses. Applying this methodology to the Scopus database of international scientific journals, the authors found out that China's share increased from 4% of all articles in 2000 to 18.6% in 2016, exceeding the US total. However, this is still an underestimate as it does not consider articles written by Chinese researchers at non-Chinese addresses and articles in Chinese language journals that are not included in the Scopus database. When these elements are considered, China's contribution accounts for 36% of the world's scientific publications. China's move to the forefront of scientific inquiry makes it a key driver of the direction of scientific and technological progress and of the knowledge-based economies of the foreseeable future.

It is evident that as universities and other higher education institutions became one of the founding blocks for a modern "knowledge-dependent economy", their roles have increased, but then so has the public interest in their functioning. Institutions of higher education are big providers of services, large employers, and receivers of significant public funds. In other words, on the one hand, higher education has become too important to be left to higher education institutions and academics alone, but, on the other hand, it must have enough institutional autonomy and respect of academic freedom in order to be able to respond to such challenges. Identifying appro-priate policies is thus a global challenge.

Efficiency of public policies

While international rankings respond to a public policy concern, also linked to New Public Management tools, which has had a profound impact on the culture of evaluation, they have also helped to reconfigure geopolitics in terms of train-ing and research. The question that needs to be asked, however, is whether the countries that spend the most on education are also the ones that get the most flattering rankings. In other words, it is about the efficiency of public spending.

In a recent paper published by *Nature* (Wagner & Jonkers, 2018), the authors analyse whether there is a relation between publication and citation for 36 nations, along with government expenditures on science. They found that, although government spending on research and development (R&D) does correlate with the number of publications produced, it does not corre-late with scientific impact, at least as assessed by citations.

In terms of papers published, the United States and China dominate as can be seen in Figure 9 by the size of the bullet point associated to each country. For papers written with international co-authors from more than one country, the United States still leads, followed by the United Kingdom, China, Germany, France and Canada. However, when the authors consid-ered this number in percentage to the total number of articles published by each country, Switzerland (42%) appears as the most connected country, followed by Belgium (38%), Singapore (37%), Austria (36%) and Denmark, the Netherlands and Sweden (all 34%). In terms of impact for international papers, Singapore tops the list, followed by the United States and then Sweden, Belgium, Switzerland and the Netherlands.

To understand the factors that could explain the impact factor of publi-cations, Wagner & Jonkers used in addition to international collaboration, scientific mobility by taking into account new researchers coming in, as well as returnees and emigrating researchers. These variables were finally used to create an index of openness. Using this new variable, the authors show that countries that are highly "open" produce high-impact research. The corre-lation between openness and citation impact was tight ($R^2 = 0.7$ according to a regression analysis) regardless of R&D spending or numbers of articles published. Thus, it appears that Public R&D funding is tied to publication output. The more money spent, the more articles produced. But it has been found that there is only a weak correlation between spending and impact. In other words, more government funds spent does not necessarily result in more citations. Countries with low openness and low impact are located in the lower-left quadrant of Figure 9. Against expectations, South Korea (which spends a higher percentage of its GDP on R&D than almost every country, including the United States) and China belong to this category.

Figure 9 – Openness and impact of research. Source: Nature, vol. 550 (32 - 33), 5 October 2017

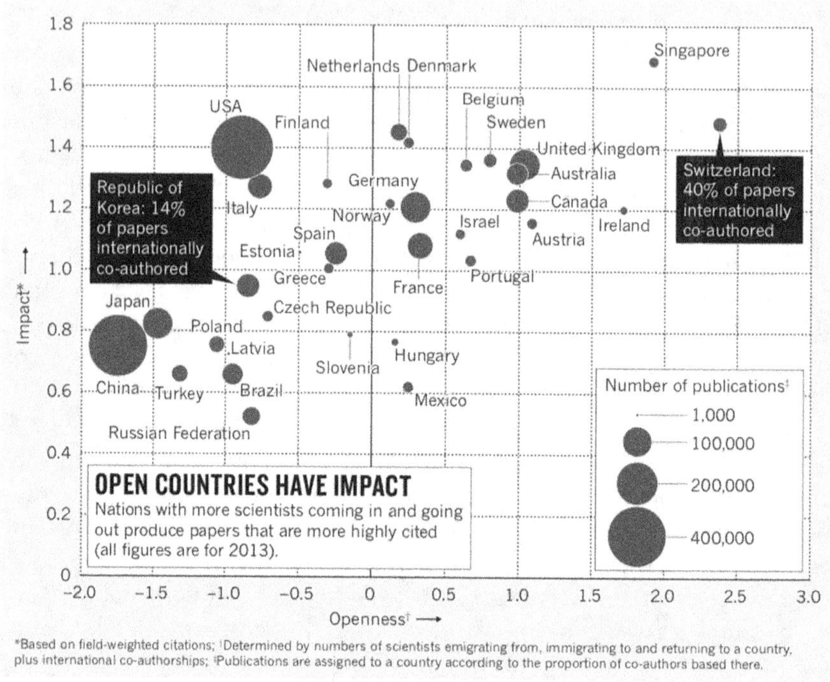

*Based on field-weighted citations; †Determined by numbers of scientists emigrating from, immigrating to and returning to a country, plus international co-authorships; ‡Publications are assigned to a country according to the proportion of co-authors based there.

Many of the countries whose research has high impact, and whose policies encourage international engagement, are from Europe. The EU has established the European Research Area (ERA) and its governments have been implementing measures to strengthen domestic research systems while also promoting both international collaboration and mobility. Analysis of citation strength shows that many European countries have greatly enhanced their impact compared with the United States. As a bloc, the EU now outperforms the United States. Both far exceed China in impact, although China's share of high-impact papers is growing rapidly.

This analysis suggests that national funding programs should whenever possible move away from policies that fund only national researchers. In the longer term, countries could benefit more by funding the best science, wherever it is, and ensuring that domestically based scientists are linked with it. Restricting the movement of researchers could be counterproductive.

In terms of training, the effectiveness of public spending on tertiary education, it may be interesting to examine the relationship between the percentage of students enrolled in high quality universities and public spending on tertiary education. Looking at Figure 10, it appears that there seems to be a relationship between the two variables that can be illustrated by an efficiency "frontier" that relates input (public expenditure) to output measured in this case through the share of students enrolled in a university ranked among the top 200 in the Shanghai ranking.

This figure shows that the United States has a relatively inefficient tertiary education system with a low proportion of students enrolled in a very good university compared to the public investment made, probably because almost all young Americans are enrolled in tertiary education. On the other hand, Italy, which has few universities ranked in the top 200, nevertheless obtains a very satisfactory result if we link it to public investment. Switzerland is close to the efficiency frontier, but could improve its performance by possibly accepting a greater concentration of its strengths among the best universities. It could be seen as the price to pay for an educational policy that has other objectives such as regional policy or linguistic diversity.

Figure 10 – Efficiency of public spending on education
(CSRE, 2019, p. 194) % of students in one of the top 200 universities
(Shanghai Ranking 2016); Education expenditure per person in tertiary
education compared to GDP per capita, 2014. Note: The curve in the graph
represents the hypothetical efficiency limit, i.e. the maximum rate of students
in one of the best universities that the expenditure considered achieves.

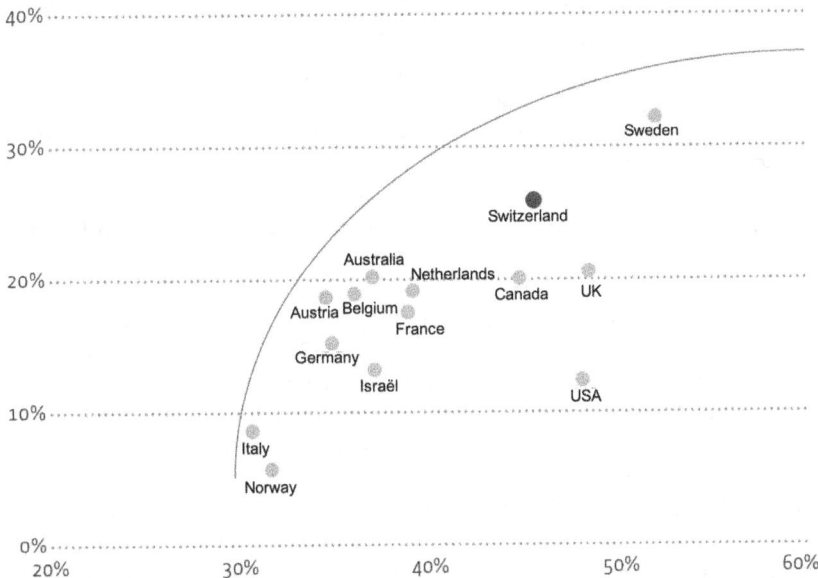

Source: <u>Data</u>: OECD, Eurostat, internet research carried out by Centre suisse de coordination pour la recherche en éducation (CSRE). Calculations: CSRE.

CONCLUSIONS

With globalization, the field of higher education has become a competitive global market where universities must attract the best talent to be recognized as the best. In this context, international rankings, which have emerged from this globalization to give a measure of university excellence, greatly influence educational policies, institutional funding and stakeholders. The funding of institutions is more and more linked to rankings and scientific production, but not really to the impact of it on society, nor do they reflect the effectiveness of a particular education system. Over the past 20 years, we have seen a change in the geopolitics of higher education, with the rise of certain regions, such as China, alongside the traditional European and North American institutions. Today, with ever-increasing mobility of talent, the upheavals we are witnessing will continue in the future and continue to modify the geopolitics of higher education. This is all the more so since, as we have been able to highlight, internationalization contributes to significantly improving the effectiveness of public policies in the field of higher education.

REFERENCES

Achard, P. & Flückiger, Y. (2016). "From MOOCs to MOORs: a Movement towards Humboldt 2.0", in Weber, L. E. & Duderstadt, J. J. (Eds), *University Priorities and Constraints*, Economica, Paris, pp. 273-283.

Centre suisse de coordination pour la recherche en éducation. (2019). "L'éducation en Suisse, rapport 2018", Aarau.

China Education Yearbook Editorial Board. (1991). China Education Yearbook 1991. People's Education Press, Beijing.

Flückiger, Y. & Raboud, D. (2018). "Une université d'excellence au cœur de la région : maillon essentiel de l'innovation et de la compétitivité", in *Le Grand Genève dans tous ses Etats*, under the direction of De V. Mottet, Slatkine, Geneva, pp. 67-85.

Marope, P. T. M., Wells, P. J. & Hazelkorn, E. (2013). *Rankings and Accountability in Higher Education: Uses and Misuses*, UNESCO, Paris.

Shields, R. (2013) *Globalization and International Education*, Bloomsbury, London.

Van Raan, T. (2005). "Challenges of Ranking in Universities", Leiden University.

Vetterli, M. & Escher G. (This volume). "Global science, global talent in the wake of nationalism and populism".

Wagner, C. S. & Jonkers, K. (2017), "Open countries have strong science", *Nature* 550, 5 October 2017, 32-33.

Xie, Q. & Freeman, R. B. (2019). "Bigger Than You Thought: China's Contribution to Scientific Publications and Its Impact on the Global Economy", *China & World Economy*, 1–27, Vol. 27, No. 1, 2019.

Table 1: Shanghai ranking weighted by the overall budget of universities (extract 50/100)

Rank	Intitution	Country	Score in Shanghai	Budget (M$)	Normalized score	Rank in Shanghai
1	Ecole Normale Superieure - Paris		2,86E+01	160	0,17875	69
2	University of Paris-Sud (Paris 11)		3,50E+01	336	0,104166667	41
3	Rockefeller University		3,66E+01	370	0,098918919	36
4	California Institute of Technology		5,73E+01	641	0,089391576	9
5	Moscow State University		2,66E+01	350	0,076	93
6	Pierre and Marie Curie University - Paris 6		3,55E+01	542	0,065498155	40
7	Technion-Israel Institute of Technology		2,66E+01	411	0,064720195	93
8	Leiden University		2,70E+01	467	0,057815846	88
9	Ghent University		2,86E+01	504	0,056746032	69
10	Erasmus University Rotterdam		2,84E+01	588	0,04829932	73
11	Stockholm University		2,82E+01	590	0,04779661	74
12	University of Oxford		6,01E+01	1336	0,04498503	7
13	McGill University		2,89E+01	646	0,044736842	67
14	Karolinska Institute		3,33E+01	760	0,043815789	44
15	University of Cambridge		7,09E+01	1643	0,043152769	3
16	Rice University		2,82E+01	680	0,041470588	74
17	University of Geneva		3,10E+01	779	0,039794608	60
18	The University of Western Australia		2,68E+01	677	0,039586411	91
19	University of California, Santa Cruz		2,60E+01	662	0,039274924	98
20	University of Washington		5,03E+01	1300	0,038692308	13
21	McMaster University		2,90E+01	752	0,03856383	66
22	Mayo Medical School		2,85E+01	750	0,038	71
23	Uppsala University		2,98E+01	800	0,03725	63
24	University of Helsinki		3,15E+01	860	0,036627907	56
25	Cardiff University		2,59E+01	724	0,035773481	99
26	University of Oslo		2,99E+01	840	0,035595238	62
27	University of Bristol		3,01E+01	869	0,034637514	61
28	University of Basel		2,63E+01	762	0,034514436	95
29	University of California, Santa Barbara		3,32E+01	1000	0,0332	45
30	Utrecht University		3,30E+01	996	0,03313253	47
31	Heidelberg University		3,48E+01	1057	0,032923368	42
32	Princeton University		6,11E+01	1926	0,03172378	6
33	University of Groningen		3,11E+01	983	0,031637843	59
34	King's College London		3,31E+01	1073	0,030848089	46
35	The Australian National University		2,61E+01	855	0,030526316	97
36	Imperial College London		4,09E+01	1381	0,02961622	27
37	The University of Edinburgh		3,70E+01	1295	0,028571429	32
38	Nagoya University		2,72E+01	1009	0,026957384	84
39	University of Copenhagen		3,85E+01	1435	0,026829268	30
40	The University of Manchester		3,61E+01	1388	0,026008646	38
41	KU Leuven		2,69E+01	1051	0,025594672	90
42	University College London		4,71E+01	1846	0,025514626	16
43	University of California, Los Angeles		5,25E+01	2063	0,025448376	12
44	Kyoto University		3,67E+01	1450	0,025310345	35
45	Georgia Institute of Technology		2,71E+01	1071	0,025303455	85
46	Vanderbilt University		3,20E+01	1274	0,025117739	52
47	Swiss Federal Institute of Technology Zurich		4,41E+01	1784	0,024719731	19
48	Carnegie Mellon University		2,77E+01	1128	0,024556738	80
49	University of California, Berkeley		6,91E+01	2819	0,024512238	5
50	University of Toronto		4,16E+01	1755	0,023703704	23

Table 2: Shanghai ranking weighted by the budget per student (extract 50/100)

Univ		Score in Shanghai	N students	Budget (M$)	Budget / student	Normalized score	Rank in Shanghai
1 University of Paris-Sud (Paris 11)		3,50E+01	31400	336	10700,63694	0,003270833	41
2 Moscow State University		2,66E+01	40000	350	8750	0,00304	93
3 Ghent University		2,86E+01	41000	504	12292,68293	0,002326587	69
4 University of Toronto		4,16E+01	88766	1755	19771,08352	0,002104083	23
5 Pierre and Marie Curie University - Paris 6		3,55E+01	31000	542	17483,87097	0,002030443	40
6 University of Washington		5,03E+01	46165	1300	28159,86137	0,00178623	13
7 Leiden University		2,70E+01	28130	467	16601,49307	0,00162636	88
8 Stockholm University		2,82E+01	34000	590	17352,94118	0,001625085	74
9 McGill University		2,89E+01	35710	646	18090,17082	0,001597553	67
10 Uppsala University		2,98E+01	42559	800	18797,43415	0,001585323	63
11 KU Leuven		2,69E+01	55484	1051	18942,39781	0,001420095	90
12 Erasmus University Rotterdam		2,84E+01	28000	588	21000	0,001352381	73
13 University of Helsinki		3,15E+01	36500	860	23561,64384	0,001336919	56
14 Monash University		2,79E+01	73807	1555	21068,46234	0,001324254	78
15 University of British Columbia		3,77E+01	61113	1755	28717,29419	0,001312798	31
16 McMaster University		2,90E+01	31265	752	24052,45482	0,001205698	66
17 The University of Queensland		3,16E+01	52329	1400	26753,80764	0,00118114	55
18 The University of Edinburgh		3,70E+01	39669	1295	32645,13852	0,0011334	32
19 Cardiff University		2,59E+01	31597	724	22913,56774	0,001130335	99
20 University of California, Los Angeles		5,25E+01	43301	2063	47643,2415	0,00110194	12
21 The University of Melbourne		3,59E+01	48000	1600	33333,33333	0,001077	39
22 University College London		4,71E+01	41539	1846	44440,16466	0,001059852	16
23 University of Oxford		6,01E+01	23195	1336	57598,62039	0,001043428	7
24 University of Copenhagen		3,85E+01	38615	1435	37161,72472	0,001036012	30
25 The University of Manchester		3,61E+01	39700	1388	34962,21662	0,001032543	38
26 University of California, Berkeley		6,91E+01	41900	2819	67279,23628	0,001027063	5
27 University of Oslo		2,99E+01	28007	840	29992,50187	0,000996916	62
28 Utrecht University		3,30E+01	30000	996	33200	0,000993976	47
29 Heidelberg University		3,48E+01	29689	1057	35602,41167	0,000977462	42
30 The University of Western Australia		2,68E+01	24327	677	27829,16101	0,000963019	91
31 University of Groningen		3,11E+01	30000	983	32766,66667	0,000949135	59
32 University of Sydney		2,75E+01	56700	1646	29029,98236	0,000947296	83
33 Technion-Israel Institute of Technology		2,66E+01	14538	411	28270,73875	0,000940902	93
34 King's College London		3,31E+01	29600	1073	36250	0,000913103	46
35 University of Bristol		3,01E+01	25024	869	34726,6624	0,000866769	61
36 University of Cambridge		7,09E+01	19660	1643	83570,70193	0,000848383	3
37 Technical University Munich		3,27E+01	40841	1721	42139,02696	0,000776003	50
38 University of Munich		3,14E+01	51420	2090	40645,66317	0,00077253	57
39 Georgia Institute of Technology		2,71E+01	29369	1071	36467,02305	0,000743137	85
40 The Australian National University		2,61E+01	23761	855	35983,33403	0,000725336	97
41 University of California, Santa Barbara		3,32E+01	21574	1000	46352,09048	0,000716257	45
42 New York University		3,86E+01	59000	3273	55474,57627	0,000695814	29
43 University of Geneva		3,10E+01	16935	779	45999,40951	0,000673922	60
44 University of California, Santa Cruz		2,60E+01	16328	662	40543,85105	0,000641281	98
45 The University of Texas at Austin		3,25E+01	51331	2658	51781,57449	0,000627636	51
46 University of Goettingen		2,63E+01	31500	1346	42730,15873	0,00061549	95
47 University of Wisconsin - Madison		3,97E+01	43820	3000	68461,88955	0,000579885	28
48 Kyoto University		3,67E+01	22657	1450	63997,88145	0,000573456	35
49 University of Zurich		3,13E+01	25542	1400	54811,68272	0,000571046	58
50 Purdue University - West Lafayette		2,80E+01	41573	2094	50369,23003	0,000555895	77

CHAPTER 2

Recruiting International Talent: Diverging National Policy Frameworks and Implications for Local and Global Prosperity

Meric S. Gertler

INTRODUCTION

R esearch universities have always thrived on the free circulation of people and ideas. So too have national economies. Many countries — most famously, the United States — have benefited from their ability to attract talented newcomers, who have gone on to perform path-breaking research, establish major commercial enterprises and generate wealth and prosperity. However, recent political shifts — including the rise of populism, nativism and protectionism — have led to significant reversals of longstanding policies in certain countries, making it harder to recruit students and talented professionals from abroad.

This chapter documents the evolving policy frameworks in a number of major international jurisdictions, noting how they are creating increasingly divergent positions with respect to the recruitment of international talent. I then explore the larger implications arising from these increasingly divergent approaches. I will emphasize the growing risk and uncertainty facing not just higher education institutions but also the broader pursuit of innovation and prosperity, and our ability to address grand societal challenges in an increasingly polarized and turbulent world.

DIVERGING APPROACHES TO INTERNATIONAL TALENT RECRUITMENT

The United States

Preliminary data for the US suggest that, in fall 2018, new enrolments of international students declined for the third consecutive year. These declines represent the only years of negative growth in the 12 years that the Institute of International Education (IIE) has tracked this metric (Baer, 2018) (Figure 1). The decline is most keenly felt in less selective colleges, master's-level and associates-level institutions, and universities in the Midwest, where 2017 saw particularly steep declines. For example, the *New York Times* reported that new international enrolment at the University of Central Missouri dropped by more than 60% in 2017 (Saul, 2018).

Figure 1 – US New International Enrolments, 2006-2017

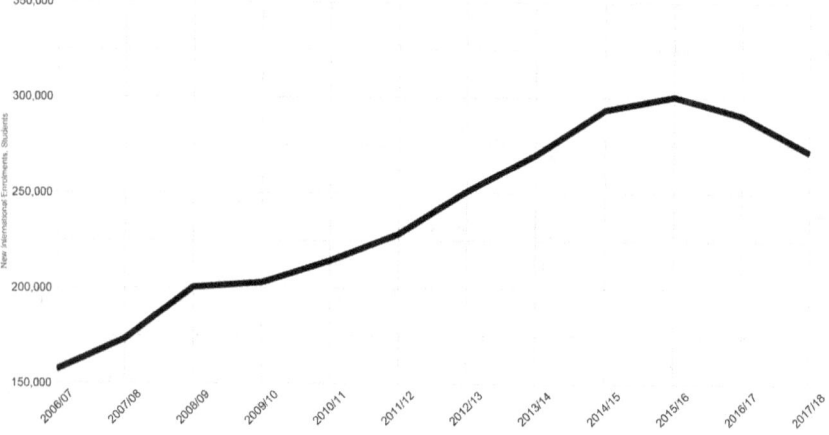

These declines have been driven, for the most part, by fewer applications from international students. In a 2018 survey of higher education admissions professionals specializing in international recruitment, 53% reported declines in applications from international students for the 2017 academic year; 45% anticipated similar declines for the 2018 academic year. (IIE subsequently reported an actual 49% drop.) Applications from China, the Middle

East and North Africa, and India showed the biggest drops: 54%, 50% and 47% respectively (Schulmann & Le, 2018). This is especially noteworthy, as China and India together account for more than half of all international students in the United States (Figure 2).

Figure 2 – International Students in the US by Place of Origin, Top 10, 2017

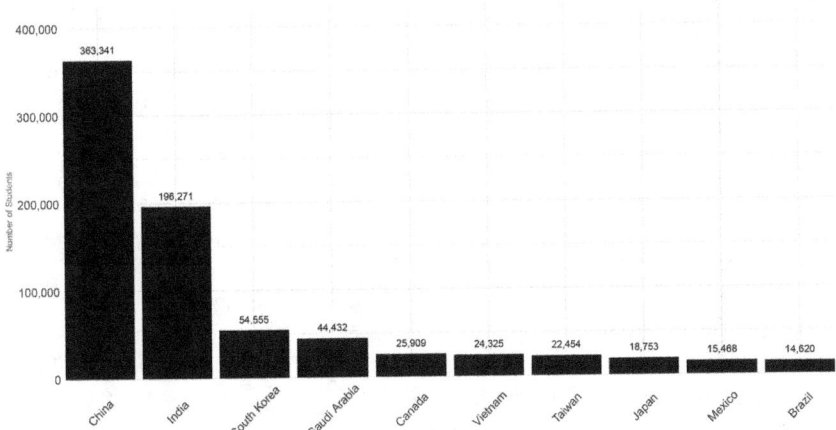

What is causing these declines? Many factors undeniably affect the international flow of students. Competition from emerging regions (most notably, China) or established regions (Canada and Australia, for example) is a significant factor, as are steadily increasing US tuition fees and the decline of scholarship programs in source countries (e.g. Brazil and Saudi Arabia).

However, the survey of university admissions professionals mentioned above provides an important insight that can be traced back to national policies on globalization. When asked: "Which, if any, of the following had a negative impact on your institution in terms of meeting international enrollment targets? (select all that apply)", the top three responses were, in order:

1. Political environment in the US (71% of respondents)
2. Increased visa delays or denials (60% of respondents)
3. Concern about securing a job or work visa in the US after studies (52% of respondents)

Aggressive and unwelcoming rhetoric from the White House has had — and is having — an impact. President Donald Trump is widely perceived to have articulated an isolationist, America-first, anti-immigration vision of

America's future. Since 2016, the Trump Administration has issued executive orders restricting immigration from certain predominantly Muslim countries; launched a campaign to build a wall along the US-Mexico border; imposed new trade tariffs; and issued calls to cut or eliminate international aid to some countries. The President has even targeted international students directly, at one point contemplating a ban on students from China while reportedly saying that "almost every [Chinese] student that comes over to this country is a spy". (Karni, 2018).

The political climate has undoubtedly had a chilling effect on international recruitment and retention, but it is difficult to quantify its impact. One proxy may be found in the analysis of visa delays or denials.

After graduation, international students typically require an H-1B temporary work visa to stay in the United States for a three-year period. They are renewable for up to six years in total, and H-1B visa holders may also apply for permanent residency (green card). The H-1B visa is also the primary mechanism by which highly educated workers holding foreign degrees in technical fields like computing, finance, engineering, mathematics, science, and medicine are admitted to work in the US. These visas are now subject to an annual cap of 85,000, which includes an allocation of 20,000 visas for workers with an advanced degree (Masters or higher) from a US academic institution.

The Trump Administration has made the application process more difficult through a series of initiatives, starting with a "Buy American, Hire American" executive order in April 2017. These initiatives have had the effect of increasing the number of H-1B visa denials and adding significant delays to employers' efforts to hire foreign, highly-educated talent (Semotiuk, 2019). For example, the proportion of visa denials has increased from 6% in 2015 to 32% through the first quarter of 2019 (24% for fiscal year 2018) (NFAP, 2019). To get a sense of how this procedure is viewed overseas, an article from the *Times of India* is illuminating. It reported on the "toughest ever H-1B visa process" amid "unprecedented scrutiny by the Trump administration" (Verma, 2018).

Strikingly, the number of F-1 student visas has declined by 44% since 2015, after years of double-digit growth. This is a product of fewer applications, but also an increased refusal rate. Approvals as a share of applications have dropped from 75% in 2015 to 65% in 2017 — and in 2017, there were 30% fewer applications than in 2015 (US Department of State, 2018). Meanwhile, for comparison, the number of H-2A temporary or seasonal agricultural work visas has continued to grow at double digit rates in recent years (Figure 3).

Figure 3 – F1 and H2A Visas Issued, 2009-2018

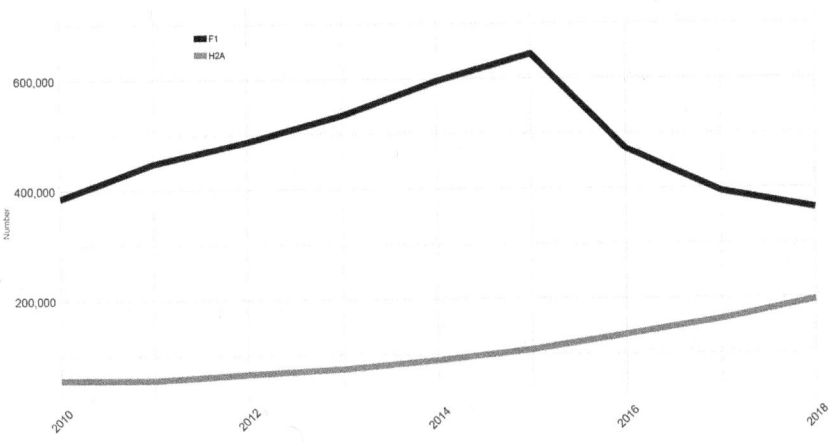

Looking ahead, this picture is likely to become even less encouraging, at least from the perspective of international students aspiring to study in the US — and those universities hoping to recruit them. The Department of Homeland Security has recently announced its intention to increase the fees for foreign students applying for F visas from $200 to $350, while exchange visitors will face an increase from $180 to $220. Sponsoring universities will face a dramatic increase in the "school certification petition fee", from $1,700 to $3,000 (Department of Homeland Security, 2019). While the changes are allegedly intended to ensure adequate resourcing to permit timely processing of visa applications, the American Council on Education, the Association of American Universities and other US higher education associations have expressed their concern that these changes "will adversely impact student and faculty exchange visitors as well as institutions of higher education... [to] reinforce a troubling message that we no longer welcome members of the international community who wish to enter our campus gates" (American Council of Education, 2019).

Taken together, these measures, accompanied by rhetoric and policies from the highest levels of the US government, reflect a purposeful retreat from international engagement. The implications, I will suggest below, may be profound.

The United Kingdom

The Brexit movement in the United Kingdom has also been interpreted as a popular retreat from international engagement and has many university leaders worrying about their future ability to attract and retain international talent. More than 17 million people, 51.9% of the electorate, voted in a national referendum on 23 June 2016 to sever Britain's ties with the European Union, a region representing half a billion people and, at just over a fifth of global GDP, the world's third-largest economy. This vote has been seen by many as a reaction against international engagement, and an apparent movement to build barriers between countries instead of bridges (Tammes, 2017). The resulting political, economic and social uncertainty in the UK has only deepened in the months and years since the referendum was held. The government's Brexit proposal has failed parliamentary votes multiple times and, at time of writing, the UK has been granted a second extension to implement an orderly Brexit. The resignation of Prime Minister Theresa May and her replacement by Boris Johnson adds further uncertainty to an already unstable political environment.

These continuing developments — and the ambivalence towards globalization underlying them — have had an interesting effect on international enrolment in UK institutions. Enrolment in UK institutions from non-UK EU nations has continued to grow – perhaps, as *The Guardian* has speculated, "a last-minute rush to study at British universities before Brexit closes the door". Enrolment from non-UK, non-EU nations has largely been flat.

I suspect we are witnessing a holding pattern, as prospective students "wait and see". EU enrolment in Britain post-Brexit (assuming the UK does eventually withdraw from the EU) will depend on the resolution of a host of thorny issues, including student mobility programs like Erasmus+, the status of the EU Framework Programme for Research and Innovation, and dozens of specific regulations concerning residency, employment, post-study work visas and professional qualifications. Recent trends have not been positive: post-study work visas were abolished by then-Home Secretary Theresa May in 2012, and the number of non-EU international student visas has been reduced.

While a recent commitment to reinstate post-study work visas is an encouraging sign, many details about Britain's future remain uncertain. If/when the UK leaves the EU, many commentators expect international enrolment in UK institutions to fall significantly. A report from the Higher Education Policy Institute and Kaplan International predicts that a rigorous Brexit that eliminates the distinction between EU and non-EU students would precipitate a 57% drop in EU students studying in the UK (Conlon *et al.*, 2017).

These same dynamics are also likely to affect the recruitment of academic talent by British universities. Currently, more than 36,000 academics employed in institutions of higher education are citizens of non-UK European countries. Though many institutions have taken steps to reassure their European and international faculty, their status in the UK remains uncertain. This fact is clearly appreciated by European institutions looking to recruit highly-qualified personnel away from UK institutions. Indeed,

Michael Arthur, Provost and President of University College London, reported to *The Guardian* that in the year following the Brexit referendum, fully 95% of his European staff had received headhunting calls from European institutions outside the UK. With the increasingly unsettled state of Brexit negotiations, retention concerns are rising at British universities.

Again, as with the United States, the implications may be profound.

Canada

By contrast, Canada's approach to recruiting international students has been aggressively positive. According to Statistics Canada, enrolment of international students in Canadian institutions of higher education jumped by more than 80% between 2010 and 2016. And the numbers continue to rise. Canada is now among the world's top five or six host countries for international students in higher education.

How has Canada accomplished this? Attracting international students is a national and provincial priority. Canada's Post-Graduation Work Permit (PGWP) program affords international students the opportunity automatically to stay in Canada for up to three years after graduation. The government has further eased the PGWP's requirements, recently extending the deadline to apply to 180 days after graduation and relaxing the requirement that students need a valid study permit at the time of application — study permits need only have been valid at some point. In addition, the Canadian Experience Class (CEC) provides a pathway to permanent residency for international students (and other non-residents) with at least one year of work experience in Canada.

The PGWP and the CEC fit into a larger international education strategy that the Trudeau government announced in its 2019 Budget. The Strategy articulates a clear goal:

Under the Strategy, the Government of Canada will work… to double the size of our international student base from 239,131 in 2011 to more than 450,000 by 2022 (without displacing Canadian students) (Government of Canada, 2019).

The parenthetical caveat is important, and I will return to it later. But the welcoming message is clear.

This same message is being reinforced by individual universities, with active recruitment programs and strategic decisions aimed at attracting highly-qualified international students. For example, in 2018 the University of Toronto reduced its tuition fees for international doctoral-stream graduate students to the much lower level that domestic students pay. Several other Canadian institutions have followed suit. In the context of rising fees in many other jurisdictions, the University of Toronto's decision resonates with price-sensitive students looking for a world-class education.

The welcoming message is getting through. According to a 2018 survey of international students (CBIE, 2018), the top three reasons international students listed for studying in Canada were:

1. The reputation of the education system in Canada
2. Canada offers a society that, in general, is tolerant and non-discriminatory
3. Canada's reputation as a safe country.

Canada has also been investing in attracting researchers and faculty who might have otherwise considered employment in the United States. The 2017 Canada 150 Research Chairs program was created expressly to recruit international world-class scholars to Canada. For example, Alán Aspuru-Guzik, a leading scholar of theoretical and computational chemistry, and of Mexican-American origin, left his tenured position at Harvard to move to the University of Toronto. To be sure, he was attracted by the opportunity to collaborate with U of T's world-class scholars in chemistry, advanced computing and machine learning. And the offer of a generously funded Canada 150 Research Chair paved the way. But, like so many of the international students surveyed by the CBIE, he also singled out Canada's welcoming inclusivity and cultural diversity as decisive factors in his decision to move to Toronto. In this light, it is noteworthy that, of the 24 Canada 150 Research Chairs, fully 13 relocated from American institutions. (It is revealing that roughly two-thirds of the University of Toronto's academic hires in the past two years have come from outside Canada — up from about 50% a few years ago.)

Moreover, the Canadian government has recently invested C$200 million to improve immigration services and make immigrating to Canada easier. Forty percent of that investment is being directed to improving the handling of work and study permits with a special emphasis on permits for foreign researchers. As Paul Davidson, president of Universities Canada, told *Times Higher Education* in April 2019: "Having this kind of concierge service for academics and their spouses will certainly help them get through our immigration process more quickly."

In addition, under Canada's Global Skills Strategy, highly educated professionals can have their visas processed within two weeks of application. For comparison, the typical processing time for an H-1B visa in the US can take between three and seven months. Employers can petition (and pay an additional US$1,225) for "Premium" processing and a 15-day turnaround. But perhaps tellingly, in 2018, Premium processing was available only during September and October. As Davidson says, the Canadian initiatives are "entirely symptomatic of a system that wants to show it is open for business" for international students and scholars (Grove, 2019). This sentiment perfectly captures the sharp contrast across the Canada-US border.

China

China has long been — and continues to be — the world's leading student exporter. Recently, however, China has focused increasing resources on hosting international students. In 2006, around 50,000 international students studied in China. By 2018, that number had grown tenfold to nearly 500,000 (Ministry of Education of the People's Republic of China, 2018). Inconsistent definitions make national comparisons difficult. But the IIE estimates that China is now the world's third-largest host country for international students.

Over roughly the same period, China has also prioritized global recruiting of highly-qualified talent. The Thousand Talents Plan was launched in 2008 to bring leading scientists, entrepreneurs and young professionals to China who "can make breakthroughs in key technologies or can enhance China's high-tech industries and emerging disciplines" (The Thousand Talents Plan, n.d.). The Plan has been the primary channel through which foreign-educated Chinese nationals with advanced degrees have been recruited back to China — some 7,000 individuals to date. Many of the recruits have worked in leading universities in the US before returning to China, lured by generous start-up packages and other perks. Some recruits have maintained their affiliation with US universities, while taking advantage of the rich research support at Chinese institutions.

As trade and global strategic tensions between China and the US spill over into research and innovation, the Plan has come under close scrutiny in the United States. Attention has been directed to Chinese-born researchers at US universities, who are now being viewed by the White House Office of Trade and Manufacturing Policy as "a primary channel for harvesting US technologies and intellectual property" (Mallapaty, 2018).

Writing in the journal *Nature*, Mallapaty notes that reports the FBI is investigating researchers involved with the Plan may be exerting a chilling effect on academic interactions between the US and China. Mallapaty argues "the threat to China-US scientific co-operation could also setback the global scientific enterprise. The two countries are the top collaborating pair in the production of high-quality scientific research worldwide, based on their joint authorship contributions of articles in the 82 journals tracked by the Nature Index."

Moreover, if the current climate deters Chinese students from studying in US universities, the consequences for research and innovation in the United States could be very significant: "About a third of all US science and engineering master's and doctorate degrees in 2015 were awarded to international students. Of the doctorate recipients on temporary visas between 1995 and 2015, some 29%, or 63,576, were from China" (Mallapaty, 2018).

LARGER SIGNIFICANCE OF THESE TRENDS

Most often, concerns about falling international student enrolment focus on financial implications. And they can be profound, both for individual institutions and for entire national systems of post-secondary education. For example, the decline in new international enrolment at the University of Central Missouri mentioned earlier will cost the institution US$14M (11% of its operating revenue), with multiplier effects in subsequent years (Saul, 2018).

A *Universities UK* study from March 2017 found that "on- and off-campus spending by international students and their visitors generated £25.8 billion in gross output for the UK economy". In the United States, the Department of Commerce estimates that foreign students contributed US$42B to the US economy in 2017 (IIE, 2019). In Canada, the number is US$15.5B, according to the most recent data (2016).

Impressive as they are, these figures do not reflect the most important contributions of international students and faculty.

Most obviously, international students — and international scholars more generally — are a tremendous source of talent to fuel economies and enrich communities. For this reason, the trends in the US and UK are a source of great concern to university presidents and innovative firms in every sector within those countries. Meanwhile, countries like Canada are reaping the benefits of openness.

For example, the University of Toronto, through its "10,000 PhDs Project", tracked the career paths of the 10,886 PhD students who graduated from that institution between 2000 and 2015. Of all international students who earned a PhD from the University of Toronto over that period, 46% are now employed in Canada, resulting "in a significant 'brain gain'." (University of Toronto, 2018).

Toronto's burgeoning AI and machine-learning ecosystem is a clear example of the propulsive force of Toronto's "brain gain". Professor Geoff Hinton came to Toronto via the UK and the US and, together with his students, has established the University of Toronto at the forefront of AI and machine learning research. His research has attracted bright students and leading scholars, who have themselves gone on to produce and attract even more talent. All of this activity has stimulated inward investment and helped create local start-ups and entirely new fields of research and development, propelling Toronto into the vanguard of this field. A case in point is Raquel Urtasun, a computer scientist and colleague of Hinton's. Her research on machine learning and computer vision induced Uber to create a large research lab in Toronto, built around Urtasun and her graduate students.

Toronto's new Vector Institute, established in March 2017, is already Canada's leading hub of artificial intelligence research, development and application — and has quickly established itself as one of the world's leaders in machine learning and deep learning. In addition to Geoff Hinton, nearly every member of the founding research faculty (including Urtasun) was recruited from abroad — including some Canadians who were repatriated after earning international degrees and/or work experience.

Evidence from the United States is similarly compelling. A 2019 analysis by *National Geographic* found that 44 of the top 100 Fortune 500 companies by revenue were founded or co-founded by immigrants or US-born children of immigrants. The list of such firms includes well-known examples like Tesla and Alphabet, but also blue-chip pillars such as Pfizer, Proctor and Gamble, Dow-Dupont, Goldman Sachs, and Bank of America (McNaughton & Nowakowski, 2019). Moreover, Saxenian (2002) has demonstrated compellingly how Silicon Valley's success was fuelled by the large numbers of local immigrant entrepreneurs. Many of these came to study at Stanford, Berkeley or other local universities, then remained in the Valley following graduation to become major players in the innovation ecosystem. Furthermore, she shows how their success has ultimately contributed to the emergence of tech clusters in India, Taiwan and elsewhere — another clear example of the broader benefits of openness and engagement.

The importance of international engagement goes beyond its implications for economic growth and local prosperity. International engagement is fundamental to innovation and understanding. Many of the grand challenges of our time are global in scope. Challenges like climate change and sustainable development, epidemics, international refugee crises, poverty and cybersecurity do not respect international borders. Solutions to these challenges are unlikely to come from scholars or research teams working in isolation. Global challenges will require global collaborations. And, as the *Nature* article cited above implies, the current isolationist climate could have perverse impacts at precisely the time when global co-operation is most sorely needed.

Moreover, I would argue that international engagement extends beyond collaborations among global institutions, to include the recruitment of international students and faculty.

As noted previously (Gertler, 2018a), the literature on creativity, collaboration and innovation emphasizes that internally diverse teams are more likely to generate innovative solutions to various problems. Teams, firms or regions collaborating under conditions of "resource heterogeneity" often perform better on creative, problem-solving or innovative tasks than those collaborating under conditions of "resource homogeneity". Bart Nooteboom and his colleagues describe the "knowledge stretching" that occurs when team members who bring different perspectives, expertise and experience to a common project interact with one another, and note how this process leads to breakthrough innovations (Nooteboom *et al.*, 2007).

This "knowledge stretching" may help explain why publications with international co-authors are disproportionately represented among many institutions' most highly cited research (Gertler, 2018a).

There may also be a local dimension to this. Just as research collaborations among different institutions around the globe produce disproportionately influential publications, so too might one might hypothesize that more diverse *local* communities of students and scholars (defined in terms of national origin) would similarly produce disproportionately influential research. Among other things, internationally recruited talent brings not only an enriching diversity of perspectives, expertise and experience, but also access to globally distributed social networks of academic and industrial colleagues. Here too, the capacity of such internationalized research teams to enhance progress by helping solve global grand challenges would seem to be considerable.

If this hypothesis is correct, it has obvious implications for international recruitment and public policy. It is also a reminder of how political rhetoric and decisions shaping immigration and talent recruitment policy, including work permits, student visas, quotas and more — can have a significant impact on both the regional and national capacity for innovation and long-term prosperity.

CONCLUDING OBSERVATIONS

It is clear from the preceding analysis that some of the world's most significant economies and centres of higher education and research are pursuing increasingly divergent paths with respect to the recruitment of international talent — students, faculty, and highly educated professionals more broadly. The consequences of this divergence are striking and significant.

For countries such as the United States and United Kingdom, recent political events, public discourse and shifts in public policy have already had a discernible effect on flows of international students, faculty and workers in technology-based sectors. By contrast, countries like Canada have clearly reaped the benefits of embracing a more open and welcoming stance. The long-term consequences of such shifts, should they prove to be enduring, could be profound.

For example, speaking at the 2019 Collision Conference in Toronto — a major global gathering of technology entrepreneurs, venture capitalists, angel investors and others — Prime Minister Justin Trudeau attributed much of the recent boom in technology-related investment and employment creation in Toronto, and Canada more broadly, to federal immigration policy for students and knowledge workers: "We're at a time where big countries around the world are closing themselves off more to immigration, at a time when Canada is realizing we need to stay open and draw in the best and the brightest from around the world" (Lindzon, 2019; CBC News, 2019). Indeed, Toronto alone created more new tech jobs in 2017 than the San Francisco Bay Area, Seattle and Washington DC *combined* (CBRE, 2018).

Meanwhile, in the US and UK, many leaders from the higher education, technology and related sectors are expressing growing concern about the impact of protectionist and isolationist pronouncements and policy stances on their viability and competitiveness. They have underscored how the creation of new barriers to international mobility is anathema to innovation and prosperity. And, as Mallapaty (2018) points out, the consequences for advances in global scholarship may also be profound.

At the same time, in both "opening" and "closing" countries, recent history demonstrates how vulnerable higher education has become in the face of major geopolitical upheaval and turbulence. Recent events in Canada provide troubling examples. A diplomatic row with Saudi Arabia resulted in all Saudi students and medical trainees in Canada being recalled (at least temporarily). Similarly, the diplomatic dispute with China over the arrest of a prominent Chinese executive for possible extradition to the United States has raised understandable concerns among university leaders in Canada, though, at the time of writing, academic interactions between the two countries remain strong.

Other domestic trends and considerations may also shape popular attitudes and political debates with respect to recruitment of international students. As I have argued elsewhere (Gertler, 2018b), the increasingly rancorous debates over access to higher education — and particularly, the growing perception that the most elite institutions remain inaccessible bastions of privilege for the select few — pose significant challenges for those who argue in favour of liberalizing the recruitment of international students.

Put simply, if domestic students are perceived to be unable to get into leading universities in their own country, is it any wonder that popular (or populist) opinion stands against increasing international enrolment? This point speaks to the importance of the parenthetical statement — "without displacing Canadian students" — included in Canada's international education strategy, as noted above. It also highlights an important but less widely appreciated connection between domestic and international policy dynamics. Our success in promoting better access to higher education for domestic students from the widest range of socioeconomic backgrounds may well have a major bearing on public opinion concerning the recruitment of talented young people from around the world to study in our leading universities.

Ultimately, the debate over the role of international talent and international recruitment is a debate about the value of openness, diversity and collaboration. In unsettled times, it is tempting to retreat inward, to build walls and play to our basest instincts. But to do so will severely undermine global prosperity, jeopardizing our ability to answer humanity's grand challenges and advance our collective well-being.

REFERENCES

American Council on Education. (2019). Open letter to Sharon Snyder, Unit Chief, Student and Exchange Visitor Program, US Immigration and Customs Enforcement, Department of Homeland Security, 17 September. Online: https://www.aau.edu/key-issues/ace-and-aau-submit-comment-department-homeland-security-visa-fees

Baer, J. (2018). "Fall 2018 International Student Enrollment Hot Topics Survey", IIE Center for Academic Mobility Research and Impact. Online: https://www.iie.org/en/Research-and-Insights/Publications/Fall-2018-International-Student-Enrollment-Hot-Topics-Survey

CBC News. (2019). "Trudeau credits immigration for Canada's growing tech sector at Toronto conference", 20 May. Online: https://www.cbc.ca/news/canada/toronto/trudeau-credits-immigration-for-canada-s-growing-tech-sector-at-toronto-conference-1.5142924

CBIE. (2018). "The Student's Voice: National Results of the 2018 CBIE International Student Survey", CBIE RESEARCH IN BRIEF, Canadian Bureau for International Education. Online: https://cbie.ca/wp-content/uploads/2018/08/Student_Voice_Report-ENG.pdf

CBRE. (2018). "Toronto rises to #4 in CBRE's annual tech talent scorecard", 25 July. Online: http://www.cbre.ca/EN/mediacentre/Pages/Toronto-Rises-to-4-in-CBRE%E2%80%99s-Annual-Tech-Talent-Scorecard.aspx

Conlon, G., Ladher R. & Halterbeck, M. (2017). "The determinants of international demand for UK higher education, Higher Education Policy Institute", Report 91. Online: https://www.hepi.ac.uk/wp-content/uploads/2017/01/Hepi-Report-91-Screen.pdf

Coughlan, S. (2018). "Australia overtaking UK for overseas students", BBC News. https://www.bbc.com/news/education-44872808

Department of Homeland Security. (2019). "Adjusting program fees for the student and exchange visitor program", Docket number ICEB-2017-0003, https://s3.amazonaws.com/public-inspection.federalregister.gov/2019-10884.pdf

Ferguson, D. (2018). "Cambridge tops the league … as Britain's most unequal city", *The Guardian*, 4 Feb. Online: https://www.theguardian.com/uk-news/2018/feb/04/cambridge-most-unequal-city-population-divide-income-disparity

Gertler, M. S. (2018a). "Global Research Collaboration: a Vital Resource in a Turbulent World", in *The Future of the University in a Polarizing World*, Weber, L. & Newby, H. (Eds.), The Association Glion Colloquium, Geneva.

Gertler, M. S. (2018b). "Higher education in turbulent times: a geographical perspective", *Journal of the British Academy*, vol. 6, pp. 239-258. Online: https://www.thebritishacademy.ac.uk/publications/higher-education-turbulent-times-geographical-perspective

Government of Canada. (2019). "Canada's International Education Strategy". Online: https://international.gc.ca/global-markets-marches-mondiaux/education/strategy-strategie.aspx?lang=eng

Grove, J. (2019). "Canadian brain gain policies are bearing fruit, says sector chief", *Times Higher Education*. Online: https://www.timeshighereducation.com/news/canadian-brain-gain-policies-are-bearing-fruit-says-sector-chief

IIE. (2019). "Economic Impact of International Students", Institute of International Education. Online: https://www.iie.org/Research-and-Insights/Open-Doors/Data/Economic-Impact-of-International-Students

Karni, A. (2018). "Trump rants behind closed doors with CEOs", *Politico*. Online: https://www.politico.com/story/2018/08/08/trump-executive-dinner-bedminster-china-766609

Leonhardt, D. (2017). "The Assault on Colleges — and the American Dream", *The New York Times*. Online: https://www.nytimes.com/2017/05/25/opinion/sunday/the-assault-on-colleges-and-the-american-dream.html

Lindzon, J. (2019). "Top takeaways from Toronto's Collision tech conference", *Globe and Mail*, 28 May. Online: https://www.theglobeandmail.com/business/article-top-takeaways-from-torontos-collision-tech-conference/

McNaughton, S. & Nowakowski, K. (2019). "Entrepreneurs with immigrant roots founded top firms", *National Geographic*. Online: https://www.nationalgeographic.com/magazine/2019/02/immigrant-entrepreneurs-founded-top-companies-data-shows/

Mallapaty, S. (2018). "China hides identities of top scientific recruits amidst growing US scrutiny", *Nature* vol 24. Online: https://www.nature.com/articles/d41586-018-07167-6

Martin, I. (2018). "Benchmarking widening participation: how should we measure and report progress?" *Higher Education Policy Institute*, Policy Note 6. Online: http://www.hepi.ac.uk/wp-content/uploads/2018/04/HEPI-Policy-Note-6-Benchmarking-widening-participation-FINAL.pdf

Ministry of Education of the People's Republic of China. (2018). "Statistical report on international students in China for 2018", Ministry of Education, People's Republic of China. Online: http://en.moe.gov.cn/news/press_releases/201904/t20190418_378586.html

National Foundation for American Policy (NFAP), (2019). "H-1B Denial Rates: Past and Present", NFAP Policy Brief. Online: https://nfap.com/wp-content/uploads/2019/04/H-1B-Denial-Rates-Past-and-Present.NFAP-Policy-Brief.April-2019.pdf

Nooteboom, B., Van Haverbeke, W., Duysters, G., Gilsing, V. & van den Oord, A. (2007). "Optimal cognitive distance and absorptive capacity", *Research Policy*, vol. 36, no. 7, pp. 1016–1034.

Saul, S. (2018). "As Flow of Foreign Students Wanes, U.S. Universities Feel the Sting", *The New York Times*, 2 January. Online: https://www.nytimes.com/2018/01/02/us/international-enrollment-drop.html

Saxenian, A. (2002). "Brain Circulation: How High-Skill Immigration Makes Everyone Better Off", *The Brookings Review*, Vol. 20 No. 1. Online: https://www.

brookings.edu/articles/brain-circulation-how-high-skill-immigration-makes-everyone-better-off/

Schulmann, P. & Le, C. (2018). Navigating a New Paradigm for International Student Recruitment. *World Education Services*, New York. Online: https://wenr.wes.org/2018/06/navigating-a-new-paradigm-for-international-student-recruitment

Semotiuk, A. J. (2019). "Recent changes to the H1B visa program and what is coming in 2019", *Forbes*, 2 January. Online: https://www.forbes.com/sites/andyjsemotiuk/2019/01/02/recent-changes-to-the-h1b-visa-program-and-what-is-coming-in-2019

Tammes, P. (2017). "Investigating Differences in Brexit-vote Among Local Authorities in the UK: An Ecological Study on Migration- and Economy-related Issues", *Sociological Research Online*. Online: https://doi.org/10.1177%2F1360780417724067

The Thousand Talents Plan. (n.d.). Online: http://www.1000plan.org/en/plan.html

The Times of India staff. (2009). "Thousands rally against racism in Melbourne", *The Times of India*, 1 June. Online: https://timesofindia.indiatimes.com/Thousands-rally-against-racism-in-Melbourne/articleshow/4599752.cms

University of Toronto. (2018). *Employed and Engaged: Career Outcomes of Our PhD Graduates, 2000-2015*, School of Graduate Studies, University of Toronto. Online: http://www.sgs.utoronto.ca/about/Pages/10,000-PhDs-Project.aspx

US Department of State. (2018). Report of the Visa Office 2018. Online: https://travel.state.gov/content/travel/en/legal/visa-law0/visa-statistics/annual-reports/report-of-the-visa-office-2018.html

Verma, J. (2018). "Toughest ever H-1B visa process begins today", *The Times of India*, 2 April. Online: https://timesofindia.indiatimes.com/india/toughest-ever-h-1b-visa-process-begins-today/articleshow/63572651.cms

CHAPTER 3

Global science, global talent in the wake of nationalism and populism

Martin Vetterli and Gérard Escher

U S President Donald Trump is part of a chilling parade of politicians [...] who have risen to prominence in the past decade by fuelling anti-immigrant sentiment. But [...] we should be grateful for what global talent has done for our economy. Since 1900, immigrants have made up one-third of US recipients of Nobel prizes in chemistry, physics, medicine and economics. Immigrants account for more than one-quarter of the approximately 110,000 patents filed in the United States each year. There are more than 1 million foreign students in US universities, representing about 5% of enrollees and providing an estimated US$39 billion annual stimulus to the economy. William R. Kerr (2018a)

Globalization is a key feature of the 21st century. The global playfield did not come naturally to Western universities, many of which were created in the 19th century and had a clear national or regional orientation. Today, the best universities — as in best-ranked universities — are also the most connected and the most international. Globalization brought along deep societal changes; today we experience a backlash which asks for tighter control of immigration and for economic protectionism. In many countries there is now a majority opinion that immigration and trade openness must be aligned with national needs. This can threaten world-class universities who run on the engine of openness. To these universities, the willingness of a country to attract foreign talent is fundamental to sustaining the quality of its national science and engineering workforces.

In this paper we first try to map the extent of the global academic talent, then we analyse recent forms and features of academic internationalization, and finally we discuss the challenges of attracting global talent today.

KEY DATA ON HIGHER EDUCATION DEMOGRAPHICS

ELIGIBILITY FOR TERTIARY EDUCATION How many young people are there available to universities? In 2015, there were 715 million people aged 18-23 globally — data in this chapter are from National Science Board (2018) or OECD (2018) — a number that will reach 800 million by 2040. Three quarters of this growth will be attributable to just nine African countries, plus Pakistan. However, our planet is aging, and the college-age population will represent just 8.2% of the total population in 2050, a three percentage point decrease from today.

STUDENT ENROLMENTS How many students will there be in the future? While predictions are always shaky, Fig. 1 presents a projection based on UNESCO numbers (Calderon, 2018). Today there are about 200 million tertiary students; there will be three times more in 25 years. However, the regions that historically dominated the student world, North America and Europe (i.e. the West), will see a mighty loss of influence. At the beginning of this century, the West still had about a quarter of total students. This number might decline to about 7.5% by 2040; the small rise of enrolments will be mostly attributable to immigration. Note also the later take off of Africa (after 2030).

Figure 1– Projected student enrolments, total number and
for selected regions. Source: Calderon (2018).

(millions of students)	2016	2030	2040
Total	215	377	594
East Asia	71	149	258
South Asia	42	91	160
North America and Western Europe	37.5	41	44
Sub-Saharan Africa	7.4	8.8	22

STEM (S&E) STUDENTS AND PHDS (National Science Board, 2018) In 2014, more than 7.5 million first university degrees were awarded in S&E worldwide. Students in India or China earned about half of those degrees, those in the European Union earned about 12%, and those in the United States earned about 10%. China and India are expected to produce 60% of young STEM-degree holders by 2030. At the doctoral level, Western research universities deliver still about half of 230,000 S&E doctoral degrees that were awarded worldwide in 2014: 73,000 degrees earned in the EU, 40,000 in the United States, 34,000 in China, 19,000 in Russia and 13,000 in India.

TOTAL NUMBERS OF MIGRANTS The proportion of migrants, 3.4% of world population in 2017 (258 million) has surprisingly changed little over the last 100 years (Pison, 2019). There were about 35 million migrants with tertiary education in the OECD in 2010/11 (OECD-UNDESA, 2013). In most countries, the emigration rates of college-educated individuals are greater than those of their less educated compatriots; Mexico and Russia are notable exceptions. One in every nine persons in Africa with a tertiary degree lived in the OECD in 2010/2011; and migrants from India, China and the Philippines accounted for one-fifth of all tertiary educated migrants in the OECD area. This implies that intra-OECD migration is (still) very high. Noteworthy: since the beginning of the century, high-skilled female migrants outnumber high-skilled male migrants (Arslan *et al.*, 2014)

INTERNATIONAL STUDENTS Overall the volume of student mobility is at an all-time high. There were 5 million international students in 2016, up from 2.1 m. in 2001 (and 1.3 m. in 1990). More internationally mobile students go to the US than to any other country (National Science Board, 2018), 19% of internationally mobile students worldwide. Other top destinations include the United Kingdom (10%), Australia (6%), France (5%), Russia (5%) and Germany (5%); these six countries host together about half of all internationally mobile students. In absolute numbers, the United States remains the top destination with about 1 million students, but its share is declining (25% in 2000, 19% in 2014 [OECD, 2017]). Of these one million students, Chinese and Indian students accounted for half. In most OECD countries, international students make up a significant part of doctorates (37% in the US; and 52% in Switzerland), reflecting the international attractiveness of research universities.

INTERNATIONAL STUDENT SITUATION IN SWITZERLAND (OFS, 2017) International students make up 19% of the student population when all types of higher education institutions are considered, and they constitute 22% of all master students and 52% of all PhDs. We are also happy to report that international students in Switzerland are on average 31 years old, and that 17.8% of them have kids. One quarter of these students came from a non-European country, one fifth does not speak a Swiss national language, and two thirds of international students are concentrated in just two areas, the Zurich or the Leman area.

INTERNATIONAL STUDENTS AT EPFL (Table 2) Over the last 20 years EPFL has become an internationally recognized polytechnic university, well placed in all international rankings, attractive to international faculty, and very attractive in terms of student fees (EPFL charges all students the same fees, about €1,000 per year). The "internationality" of EPFL is higher than the national average, Swiss students are a minority at all levels of study, and the share of international students has increased sharply in a short time.

At bachelor and master levels we recruit "glocal" students, i.e. international students from nearby France; the PhD level is truly international. EPFL views internationality as a measure of success.

Figure 2– Students at EPFL, segmented according to their previous diploma.

Previous diploma	Bachelor			Master			PhD		
	2005	2010	2018	2005	2010	2018	2005	2010	2018
Switzerland	72%	64%	49%	68%	55%	40%	35%	35%	32%
France	13%	25%	38%	12%	16%	33%	12%	7%	8%
Rest of Europe	8%	7%	8%	9%	14%	14%	32%	38%	39%
Asia	2%	1%	1%	4%	9%	7%	11%	13%	15%
Others	5%	3%	4%	7%	6%	5%	9%	8%	6%

OUTLOOK OF HIGH SKILLED MOBILITY In the end, host countries may end up with high concentrations of high-skilled immigrants (Stephan et al., 2013); for example, immigrants account for some 57% of scientists residing in Switzerland, and 38% in the United States. Strong reliance on foreign talent is therefore not a sign of scientific weakness, on the contrary. In comparison, in India, Italy and Brazil less than 4% of the doctoral or postdoctoral-trained workforce is foreign-born.

INTERNATIONAL STUDENTS IN THE 21ST CENTURY

Over the past 40 years internationalization of higher education has taken several forms and accents. For a long time, internationalization was primarily focused on development, cooperation and aid. Then, particularly in Europe, the focus shifted from aid to exchange of students and curriculum development. We analyse here the developments since 2000, following closely the "three waves" segmentation proposed by (Choudaha, 2017). The underlying drivers and characteristics of these three waves suggest that academic institutions will be under increasing financial and competitive pressure to attract and retain international students.

Wave I

Wave I (1990s-2005) was shaped by the increasing demand for talented students in STEM fields, pushed by demand in biotechnology and information and communication technology; Europe was building the European Higher

Education Area, an initiative to create some coherence in higher education and to foster student mobility within Europe. Also, the terrorist attacks of 2001 made it harder for many students to enrol in American research universities. Towards the end of Wave I, five of the top 10 destination countries were in Europe (the UK, France, Italy, Austria and Switzerland). China became an important source country, with many Chinese students moving to Japan or South Korea. International students in this wave were more likely to be academically prepared in science, choosing the best universities but dependent on financial aid and scholarship from the hosting institutions (Choudaha, 2017).

ERASMUS A continuing success story from this era is the ERASMUS Program (De Wit, 2013), initiated by the European Commission in 1987. To date about nine million students in Europe have profited from this mobility program. Its budget will even double for the next funding period (FP9), to €30 billion, and it aims to internationalize about 12 million students (and apprentices) (European Commission, 2019).

Wave II

Wave II (2006–2013) Wave II was shaped by the global financial recession which triggered financial motivations for recruiting international students, as they were severe budget cuts in the higher education sector in many countries around the world. The narrative of Wave I of "attracting global talent" changed to "recruiting international students" in Wave II (Choudaha, 2017). Interest in recruiting foreign students grew as their tuition fees were often higher than for national students. In the US in particular, there was a dramatic growth of self-funded Chinese students and of government-funded Saudi students. Most students in this wave concentrated in business studies, especially at the undergraduate level.

In the UK predominantly, but also in continental Europe (Denmark, Sweden, The Netherlands), moves occurred or were planned for higher fees for international students from outside the EU. And, "against all expectations" (De Wit, 2013), it has been surprising to see that this did not result in a decrease of international students but in a substantial increase, following the principle "what you have to pay much for must be of good value" (De Wit, 2013), making the United Kingdom the number 2 and Australia the number 5 countries in receiving international students.

TEACHING IN ENGLISH In Europe there was about a ten-fold increase in masters programs taught in English (8,100 in 2014, up from 725 in 2001, [Benson & Griffith, 2018]), reflecting the will to serve international students. And if our experience at EPFL helps, teaching in English not only serves to attract students, but also helps local students to get out into the world.

Wave III

Wave III is being shaped by a combination of three major events (again we follow Choudaha [2017]). First, the economic slowdown in China is decelerating the growth of Chinese students going abroad. The second major event is Brexit and the third is the election of Donald Trump. The US and the UK are both top destination countries for international students and both events have strong anti-immigration tones.

In parallel, many countries detect skills gaps (due to aging of the population) prompting policies that align migration programs with the economic needs of the country, in part through international students. Retaining talent in line with the needs of the country is a dominant policy of wave III.

FOREIGN ENROLMENT FALLS IN THE US (Nature Careers, 2019). A first consequence, and a first in recent history, is the small decrease in the number of international students enrolling in US graduate programs. The US Council of Graduate Schools (CGS) reports that a first decrease of application from prospective international graduate students occurred in autumn 2017 (3.7% decrease), followed one year later by another 4% decrease. Somewhat unsurprisingly, substantial declines were noted for Iran, the Middle East, Europe (-13%) and India.

CHINESE ENROLMENT IN THE UK INCREASES Illustrating the race to attract students, while the enrolment of Chinese students is stalling in the US, the UK signals a huge increase of Chinese international students (Weale, 2019). Ten years ago, 45,000 Chinese students were enrolled in UK universities; today there are 130,000, and rising. Manchester University for instance has about 5,000 Chinese students for a total of 40,000. Is there thus a bright side to Brexit, meaning a competitive edge for the UK in recruiting non-EU students?

THE GROWTH OF GLOCAL STUDENTS: another consequence of a disturbed international environment is the rise of glocal students, students that aspire to gain a global experience, but at local cost (Choudaha, 2017). In the OECD, 850,000 mobile foreign students (i.e. about one out of five) come from a bordering country. Regional migrations are paramount in Asia. About one third of the 1 million mobile students in East Asia moved within the region (OECD, 2018). EPFL experienced a fantastic rise in international students from neighboring France (see Table 2).

TRANSNATIONAL EDUCATION Cross-border delivery of education (offshore education) is another trend of Wave III. The underlying assumption is "if they do not come to us, why don't we go to them". The largest exporters of branch campuses (C-BERT, 2017) were the United States (109 branch campuses), the United Kingdom (45), France (31), Russia (22) and Australia (21). The largest importers of branch campuses were China (38 branch campuses) and the Gulf states (77).

INTERNATIONAL SCIENTISTS SUSPECTED AS SPIES Raising economic tensions between countries may create a climate of suspicion directly affecting foreign scientists. The most publicized examples (e.g. Facher, 2018) are from Chinese scientists working in the US or Europe; this should not imply that Chinese scientists are particularly prone to academic espionage. Liu Ruopeng, a Chinese researcher working at Duke University who was accused of stealing information used to develop a so-called "invisibility cloak" on behalf of the Chinese government between 2006 and 2009, by running a "shadow lab" in his home country while conducting government-funded research in the US. Or Chinese student Huang Xianjun, a PhD Student in Graphene material science at the University of Manchester, one of the estimated 2500 scientists chosen by the Chinese military to study abroad under the program doing "Picking flowers in foreign lands to make honey in China". And three researchers have been ousted from MD Anderson Cancer Center (Ackerman, 2019) because NIH discovered they disclosed information about confidential grants to people with ties to foreign governments during the peer-review process. The global science enterprise, built on trust and exchange, appears to be totally unprepared for this.

TODAY/CONSEQUENCES

Immigration and Universities

IMMIGRATION AND NATIONAL PRIORITIES Countries select many of their immigrants in accordance with clearly articulated economic criteria to maintain public confidence in the governance of migration. Selection systems of immigrants, including students and academics that respond to a country's labour-market needs become the gold standard. For universities the challenge is to accept the link to local labour-market needs without losing the mission of educating the students "of tomorrow and not just of today".

One might question why high-skilled migration should ever be restricted (Pekkala et al., 2017). The primary economic arguments centre on possible adverse wage and employment effects on skilled native workers (Pekkala et al., 2016). As universities both host high-skilled immigrants AND are educating skilled native workers, they are at the heart of the discussion.

THE UNIVERSITY IMMIGRATION PATHWAY (We follow Kerr [2018b] for the argumentation). Many "skilled immigrants" arrive with only "raw talent and ambition" with the aim of improving their life through formal schooling. Universities and colleges are important gatekeepers through their selection of individuals, as student visas and student circulation are often unlimited, as exemplified by the F1 (student) or J1 (exchange visitor) visas in the US. About 700,000 such visas were delivered in 2015 alone. Such visas do not offer long-term employment, but graduates often get hired by local firms: nearly half of the new H-1B working visas in 2014 went to applicants already in the country, notably from these school-to-work transitions. In addition, temporary work visas are extended for STEM students, to 36 months after graduation by the Obama administration (with subsequent restrictions by the later administration). The university pathway has also become more important as PhD students more frequently enter the private sector (Langin, 2019); in the US in 2017, private sector employment of PhDs (42%) was now nearly on par with educational institutions (43%).

LONG-TERM STAY RATES In this new framework, the competitive economic advantage in attracting foreign students is fully realized only when these individuals stay to work after graduation. Stay rates are generally high; for PhD recipients they were (in 2015) 70% both at the 5- and 10-year stay rates in the US. The percentage of new STEM doctorates from China and India — the two top countries of origin — with definite plans to stay in the US has declined over the past decade to about 50%, as these nationals either feel unwelcome or their country of origin has built their own innovation capacity. In Switzerland (OFS, 2017), about one third of foreign graduates have left the country one year after graduation. Most move back to their country of origin (mostly France and Germany); then about one quarter of the returnees, while leaving abroad, will work in Switzerland as *frontaliers*. The total "keep rate" of Switzerland regarding foreign graduates is therefore about 75%, an excellent score; the score is less stellar for extra-European graduates (Waltersperger & Donzé, 2019).

LIMITING FOREIGN STUDENTS For the time being, Swiss universities — among many European universities — do not charge full cost to foreign students, and therefore discussions flare up regularly to submit international students to quotas. The possibility of limiting the number of foreign students at EPFL or ETH Zurich was effected into law in 2016. If these universities want to restrict access of foreign students, they have to demonstrate that the influx of foreign students (EU or non-EU) would exceed their "capacity". Except for the medical curriculum at ETH Zurich, no restrictions have been effected so far. On the contrary, at EPFL international students are seen as a sign of attractivity and, rather than establishing quotas, EPFL intends to invest in teaching, hire more professors and build more on-line modules. EPFL applies a selection (but not a quota) on French students who need a final note of 16/20 and a *mention très bien* on their baccalaureate to apply to EPFL.

RE-EVALUATING BRAIN DRAIN Emigration increases with development, because the proportion of college graduates in the native population increases and it is precisely this group that has highest propensity to emigrate abroad (Dao *et al.*, 2018). The number of tertiary-educated African migrants in OECD increased dramatically by 80% between 2000 and 2010, but the emigration rate went down. This is explained by the almost doubling of the population with tertiary education in Africa for the period. Skilled migration does not necessarily lead to a brain drain. The positive effects of skilled migration can come in the form of remittances and knowledge exchange through professional networks. In science, an effective "ethnic" network is at work, with the potential of delivering knowledge spillovers to origin countries (Tejada & Bolay, 2015). According to the World Bank, officially recorded remittances to developing countries amounted to US$414 billion in 2013. This is about three times greater than official development assistance! Medical Brain Drain however is the most problematic aspect of brain drain and is unsolved.

FULLY CAPITALIZE ON OUR OWN CITIZENS Attracting the best students is not incompatible with building a strong national "STEM workforce"; as the National Science Board (2018) puts it: "Governments and businesses should expand their investments in community and technical colleges, which continue to provide individuals with on-ramps into skilled technical careers, as well as opportunities for skill renewal and development for workers at all education levels throughout their careers." Switzerland is lucky to have a dual system of high-quality research universities and professional apprenticeships. This differentiated R&D system, with solid national or local roots for skill development creates a "vertically integrated innovation system", which leaves room for research universities to concentrate on attracting global talent and playing the global competition.

Global Talent

THE RACE TO ATTRACT TALENTS IS LIKELY TO GET TOUGHER IN THE FUTURE Global demographics, regional developments, changing student enrolments described above make it likely that there will be, in the coming years, increased competition for global talent.

In times of growing protectionism, some governments are paradoxically active in promoting mobility. Internationalization is seen as a means to increase soft power, to infrastructure capacity-building, to drive up research and teaching standards, or to solve workforce weaknesses with an aging population. An illustration is China's activities in Africa (Benson *et al.*, 2017), with 70,000 scholarships over 8 years, 40,000 training opportunities in Chinese companies, the establishment of 46 Confucius institutes over 30 countries in Africa. Canada aims for 450,000 international students by 2022 (312,000 in 2017), with a point system that advantages applicants for residency for those whose degrees were obtained in the country; Germany aimed at 350,000 students by 2020, the target was already reached in 2017, with an 18 month-time span to find employment after graduation to retain the students. China aims at 500,000 international students by 2020 (442,000 in 2016), with internships and smoother pathways to residency permits. Japan wants to boost its international student force to 300,000 by 2020, from 180,000 in 2017 with targeted recruitment, subsidized company internships, job search assistance and streamlined visas. (All examples are from [Ilieva *et al.*, 2017]).

EXCELLENCE ATTRACTS EXCELLENCE In the end, the true engine of international mobility of students is quality of teaching and excellence of research. Global university rankings are the iconic manifestation of trying to measure excellence in a truly global way. Some countries could rest on "Ivy Leagues", some others created excellence initiatives to help shape world class universities. China's strategy (C-9, the Chinese "Ivy league") (OECD, 2017) was to dedicate important resources (C-9 = 10% of whole budget) to a few universities, and has now six universities in the top 200 Times Higher Education ranking (up from two in 2011). Russia and India are also moving up the ranks. Undoubtedly students — or their sponsors — are consulting these rankings.

There are other ways to check for excellence. In the coming decades, Europe, and probably North America, will no longer be able to attract new students massively. But we must play on quality. In research, we'd move away from quantity, i.e. publication numbers, and concentrate on quality, i.e. top-cited publications. Here Switzerland is ranked first, followed by The Netherlands, Denmark, US and Great Britain. By analogy we'd deploy the best graduate programs. Our funding systems, our international connections should be designed to be able to maintain scientific quality, which in the end is what attracts the brightest students.

World class universities can become similar through the unifying pressure from key performance indicators of the global rankings, or they can converge quietly (by use of English, and converging PhD training with doctoral schools). Or the transformation can be more profound, by creating a truly global science through open science (Henry & Vetterli, 2018) and open enrolment, living by "association by the best and participation by all" (Beddington, 2019).

Let our slogan be: #youarewelcomehere!

REFERENCES

Ackerman, T. (2019). "MD Anderson ousts 3 scientists over concerns about Chinese conflicts of interest". *Houston Chronicle*, 20 April.

Arslan, C. *et al.* (2014). "A New Profile of Migrants in the Aftermath of the Recent Economic Crisis". *OECD Social, Employment and Migration Working Papers*, No. 160, OECD Publishing. Online: http://dx.doi.org/10.1787/5jxt2t3nnjr5-en.

Beddington, J. (2019). "Keep science on the horizon". *Science*, 363 (6428), 671. Online : https://doi.org/10.1126/science.aaw9290.

Benson, K. & Griffith, L. (2017). "International Trends in Higher Education 2016–17", University of Oxford. doi:http://www.ox.ac.uk/sites/files/oxford/trends in globalisation_WEB.pdf

C-BERT, Cross Border Education Research Team. (2017). Branch Campus Listing, updated 20 January 2017. Online: http://cbert.org/wp-content/uploads/2017/01/Branch-Campus-Listing-Updated-1-20-2017.xlsx.

Calderon, A. J. (2018). "Massification of Higher Education Revisited", RMIT University, Melbourne, Australia. Online: https://doi.org/ 10.1007/978-981-13-0248-0.

Choudaha, R. (2017). "Three waves of international student mobility (1999–2020)". *Studies in Higher Education*, 42(5), pp. 825–832. Online: https://doi.org/10.1080/03075079.2017.1293872.

Dao, T. H., Docquier, F., Parsons, C. & Peri, G. (2018). "Migration and development: Dissecting the anatomy of the mobility transition". *Journal of Development Economics* 132 pp. 88–101. https://doi.org/10.1016/j.jdeveco.2017.12.003.

De Wit, H. (2013). "Internationalisation of higher education, an introduction on the why, how and what". In *An Introduction to Higher Education Internationalization*, Via e Pensiero, Milan, pp. 13-45.

European Commission. (2019). Erasmus+ Factsheet. Online: https://ec.europa.eu/programmes/erasmus-plus/resources/documents/erasmus-annual-report-overview-factsheets_en.

Facher, L. (2018). "NIH Report scrutinizes role of China in the theft of US scientific research". Doi:statnews.com/2018/12/13/nih-report-scrutinizes-role-of-china-in-theft-of-u-s-scientific-research.

Franzoni, C., Scellato, G. & Stephan, P. (2012). "Foreign Born Scientists: Mobility Patterns For Sixteen Countries". *National Bureau Of Economic Research*, Working Paper 18067. Online: http://www.nber.org/papers/w18067.

Henry, L. & Vetterli, M. (2018). "Open Science, A global enterprise", in *The Future of the University in a Polarizing World*, Weber L. & Newby H., Eds, Association Glion Colloquium, pp. 71-81.

Ilieva, J., Killingley, P., Tsiligiris, V., Peak, M. & British Council. (2017). "The Shape of Global Higher Education: International Mobility of Students, Research and Education Provision". *British Council*, 2, 40. Online: https://www.britishcouncil.org/sites/default/files/h002_transnational_education_tne_ihe_report_final_web_2.pdf

Kerr, W. R. (2018a). "America, don't throw global talent away". *Nature*, 563, 445.

Kerr, W. R. (2018b). *The Gift of Global Talent: How Migration Shapes Business, Economy & Society*. Stanford Business Books, Redwood City, CA.

Langin, K. (2019). "In a first, US private sector employs nearly as many PhDs as schools do". *Science Careers*, 12 Mar. doi:10.1126/science.caredit.aax3138.

National Science Board. (2018). "Science and Engineering Indicators 2018". NSB-2018-1. National Science Foundation, Alexandria, VA. Doi: https://www.nsf.gov/statistics/indicators/.

Nature Careers. (2019). "Foreign enrolment falls". *Nature*, 565, 389.

OECD-UNDESA. (2013). "World Migration in Figures". OECD-Undesa, (October), 1–6. Online: https://doi.org/10.1080/12538078.2004.10516022.

OECD. (2017). "OECD Science, Technology and Industry Scoreboard 2017: The digital transformation", OECD Publishing, Paris. Online: http://dx.doi.org/10.1787/9789264268821-en.

OECD. (2018). "Education at a Glance 2018: OECD Indicators", OECD Publishing, Paris. Online: http://dx.doi.org/10.1787/eag-2018-en

OFS — Office fédéral de la statistique. (2017). Diplômés des hautes écoles issus de la migration — Intégration sur le marché du travail et émigration en 2015. Neuchâtel. Publication No. 1650-1500.

Pekkala, K. S., Kerr, W., Özden, C. & Parsons, C. (2016). "Global talents flows", *Journal of Economic Perspectives*, 30, 4, pp. 83-106. Online: http://dx.doi.org/10.1257/jep.30.4.83.

Pekkala, K. S., Kerr, W. R., Ozden, C. & Parsons, C. R. (2017). High-Skilled Migration and Agglomeration. *Annual Review of Economics*. Online: https://doi.org/10.1146/annurev-economics-063016-103705.

Pison, G. (2019). "Le nombre et la part des immigrés dans la population : comparaisons internationales". *Populations & Sociétés* 563, 1-4.

Stephan, P., Franzoni, C. & Scellato, G. (2013). "Choice of country by the foreign born for PhD and postdoctoral study: a sixteen-country perspective". National Bureau of Economic Research, Working Paper 18809. Online: http://www.nber.org/papers/w18809.

Tejada, G. & Bolay, J.-C. (2015). *Scientific diasporas as development partners: skilled migrants from Colombia, India and South Africa in Switzerland*. Lange, Bern.

Waltersperger, L. & Donzé, R. (2019). "Der Ingenieur, den keiner will". *NZZ am Sonntag*, 11 Aug 2019, p. 10.

Weale S. (2019). "Union européenne : les étudiants chinois affluent dans les universités britanniques". *Le Monde*, 27 May.

CHAPTER 4

Higher education: the curious case of Australia

Michael Spence

INTRODUCTION

At least since the European Middle Ages, universities have been at once both intensely local and international institutions. They are highly located in their particular social and cultural milieux, not least subject to the vicissitudes of local political conditions and sometimes control, and usually also part of an international network of research and people to people contact.

This dual outlook is not without its pressures. Internationally, it can seem as if there is some idealized hypostasis that is the modern university, the performance of which can be measured by leagues tables of various kinds. There is some truth in this. The normative dominance of the self-governing, comprehensive, research and teaching university, the corporate life of which is characterized by a commitment to some notion of academic freedom and the unfettered pursuit of ideas, means that many of the world's great universities are more or less recognizable as such across sometimes quite profound cultural difference. It also means that at meetings of university presidents from very different cultural and political contexts, there is a camaraderie in the challenges of running an institution of just such a type. But local social, political, policy and funding pressures often take profoundly different forms in different parts of the world. Navigating what it means to be both locally rooted and internationally connected can be quite different in the United States, in China, or in Australia.

On the surface, Australia's higher education sector may appear to be in the midst of an extended period of growth and success as part of an economy now in its 27th year of consecutive growth, a feat unparalleled across the OECD. Despite Australia's relatively small size and geographic isolation, our top universities perform strongly in international rankings and we are the second most popular destination for international students globally, behind only the US. However, a closer examination of our sector's situation reveals some fundamental structural weaknesses in the way Australia supports higher education and research.

Our success in international education has created challenges in two distinct but related areas. First, our universities are increasingly over-reliant on revenue from international student fees. Second, the students we attract are drawn from a small number of countries and tend to study in a narrow range of courses. This creates a "double concentration risk", which has both financial and educational consequences.

Australia's recent experience provides a cautionary tale for those in other jurisdictions looking rapidly to expand their international education activities. This paper discusses the curious case of Australia's higher education sector through the example of the University of Sydney.

THE AUSTRALIAN ECONOMIC MIRACLE

In late 2018 the *Economist* ran a feature about the Australian economic miracle of 27 consecutive years without a recession. Except for a passing reference to how one university helped the economic resilience of a regional economy after the closure of a major factory during the Global Financial Crisis, the article was silent on the role that our higher education sector has played in supporting this run of internationally unprecedented continuous economic growth. As the *Economist* noted: "The last time Australia suffered a recession, the Soviet Union still existed and the internet did not. An American-led force had just liberated Kuwait, and almost half the world's current population had not yet been born. Unlike most of its region, Australia was left unscathed by the Asia crash of 1997. Unlike most of the developed world, it shrugged off the global financial crisis, and unlike most commodity-exporting countries, it weathered the resources bust too. No other rich country has ever managed to grow so steadily for so long. By that measure Australia boasts the world's most successful economy." (McBride, 2018).

When Australia became an independent nation in 1901, the economy was heavily based around primary industry — agriculture, mining and manufacturing — and just 0.1% of the population attended university. Today, around 1.1 million Australians are enrolled in higher education — more than 4% of the population — while our economy is around 80% services-based: in areas such as health, education, community and personal services, finance, engineering, information technology, software design, telecommunications and tourism — all of which require higher education qualifications. The Reserve Bank of Australia has noted that the bulk of Australia's newest jobs, more than 3.5 million, have been created in these sectors since the early 1990s, compared to just over 500,000 in the goods-producing industries. While Australia's universities have played an important role providing the highly educated and skilled people our expanding and transforming economy has demanded over the last quarter century, our economic success has been due largely to our proximity to Asia and its rapidly growing demand for our exports. Education alone is now our largest service export earner worth A$34 billion (AUD) a year in revenue, ahead of tourism and third in overall value within our wider economy.

WHAT MAKES AUSTRALIAN HIGHER EDUCATION CURIOUS

The challenge of balancing the local and international in a modern research-intensive university takes a particular shape in Australia. In research, Australia shares the challenge of any small jurisdiction: at least some of our research needs to address local issues that are rarely of interest overseas (such as the history of New South Wales, or the habitat of the bilby), but our research must also be engaged in the global research conversation. Outside areas of purely local interest, it is more crucial than in some larger jurisdictions that our work be jointly authored with Northern Hemisphere authors if it is to gain an international readership. In the area of research, the local and the international need to find an appropriate balance.

But it is in the context of our work as educators that the challenge of bringing the local and international together takes on a distinctive characteristic. This distinctiveness can be seen in the Australian higher education funding model, our size and breadth of disciplines, and the international composition of our student mix. They, in turn, generate this "double concentration risk".

Educating Australian students is, with some exceptions, more or less a break-even activity, while funding for university research falls well short of meeting the full cost of supporting that activity. At the University of Sydney, for example, we estimate that for every dollar of externally funded research, we need to find at least another dollar from other sources. Australian research-intensive universities are, by international standards, not well-endowed, and essentially operate from our profit and loss statements. At Sydney, we have just completed Australia's first A$1 billion fundraising campaign, but it will take quite some time to accumulate the kind of endowment that can sustainably maintain our status as a world-class research university.

Given the current inadequacy of domestic tuition fees, research funding and philanthropy in Australia, the only other source of revenue available is fees from international students.

This funding model forces our institutions to become relatively large and comprehensive by international standards in order to be internationally competitive. Five of our top-performing research-intensive universities have well in excess of 40,000 students, more than 5,000 staff and budgets exceeding A$2 billion. All are still growing. At my university, we have around 70,000 enrolled students (52,000 Equivalent Full-Time Student Load or "EFTSL"), split fairly evenly between undergraduates and postgraduates. Like most other Australian research universities, we are unusually comprehensive and offer programs in both research-intensive disciplines, such as Physics, and less research-intensive disciplines, such as Speech Therapy. Looking through the top 200 universities in the THE Rankings, we find that the University of Sydney is, by field of education, perhaps the most comprehensive in the world. Admittedly, this breadth may disadvantage us in international ranking metrics, but it also provides tremendous opportunities for conducting the multidisciplinary work necessary for fields such as sustainable development.

Perhaps the most striking feature of Australian higher education today is the unusually international composition of our student body, a situation that has rapidly developed over only the last 30 years. In 1991 there were fewer than 100,000 international students studying in Australia across all levels of education, generating around A$1.2 billion in export earnings. By 2005 this had grown to 350,000 students and A$10 billion and it now stands at more than 700,000 students generating more than A$34 billion.

At Sydney, we currently have more than 25,000 international students. They account for around 40% of our enrolments, are drawn from 132 countries and are roughly equally spread between undergraduate and postgraduate courses. The revenue from their fees is our single largest source of income, accounting for more than a third of our budget. This is typical of the Group of Eight Australian research-intensive universities where international students make up between 23 to 51% of their students.

These students are drawn overwhelmingly from one region. Students from China (excluding the Special Administrative Regions and Taiwan) represent over 65% of my University's international enrolments. This dwarfs the next highest country of origin, the US, which accounts for about 5% of our international student body, mainly through semester-length exchange programs. International students are heavily clustered in a narrow range of courses, with 53% of the total international coursework cohort enrolled in a degree in business and economics or engineering and technology. At an undergraduate level, approximately one in five international students are enrolled in business or economics courses, and one in seven in engineering or technology.

The concentration is even higher at the graduate level, with 55% enrolled in business or economics, and 15% in engineering or technology.

FINANCIAL RISKS

Australian universities' dependence on international students drawn disproportionately from one country and into business and technology faculties places our institutions in a potentially precarious situation. While there are predictions that global demand for international education will continue to grow strongly until at least 2030, these predictions assume continuing global economic growth and a stable geopolitical environment, which is by no means assured. A future economic downturn or worsening tension between superpowers around trade and regional security would impact student mobility globally, including Australia.

Australia has a specific vulnerability in the form of our exposure to China, which has been investing heavily in its own higher education sector and research capability in strategic areas. The results of this investment are already evident, with Chinese institutions rising rapidly in all major rankings. While we hope and work to ensure that Australia maintains its share of the international student market, it would be misguided to assume that this will continue indefinitely, given the rise of China and competition from other popular countries such as the US and the UK, and emerging destinations in Asia.

Beyond the higher education sphere, our specific reliance on Chinese international students is a sensitive question in the present geopolitical climate. If tensions between China and the US continue to rise, Australia will find itself in an unenviable position caught between our largest trading partner and our principal military and intelligence ally. Australian universities have always had an important role in building links and understanding between Australia and China. My own University has been working with researchers in modern China since the 1960s, received its first students from that country in 1979, has over 260 staff who work on issues facing the People's Republic, and deep education and research links across the country. Our China Studies Centre and our Confucius Institute both have important, though distinct, missions in public education. We will continue to build and foster strong links between people and institutions across national borders that we hope will weather the ups and downs of political relations, but this is a delicate balancing act.

To manage the concentration risks we face, it is essential that we diversify our international student cohort. At Sydney we are beginning to engage more strategically with India through investment in brand awareness marketing, nurturing agent and partner institution relationships, and leveraging undergraduate pathways. This is a challenge given the traditionally very strong preference of Indian students for study in the US and the UK, but we are beginning to see a shift in attitudes. We are also renewing our focus on South East Asia, in which region our considerable academic expertise has not been matched by successful student recruitment. Moreover, we are working to accelerate a shift in the choices of course that international students make (a shift we have been seeing for some time) to courses across the University beyond the traditional destinations in business and technology.

MAINTAINING EDUCATIONAL QUALITY

The socio-cultural benefits of having a large international student group are significant. International students contribute to the cultural diversity of Australia's cities and the experience of domestic students in the classroom and on campus. Australia is a highly multicultural country. In Sydney, for example, 36% of the population was born outside Australia and nearly 40% of Sydneysiders speak a language other than English at home. At the University of Sydney, almost 40% of our 2018 student cohort are of a non-English-speaking background. In this context it makes sense that Sydney should be a global education hub.

However, the concentration of international students in a limited range of courses creates challenges and risks to the quality of the experience we can provide for all students. Australian universities must grapple with how to maintain a quality student cohort where students can meet and form lasting networks with people from different backgrounds. We must equip our staff to teach and our students to learn effectively in multi-cultural, multi-lingual classes (or in some cases effectively mono-cultural classes in which the dominant culture is not that of the Australian community at large). We must ensure that our international students have opportunities for rich engagement with the host culture, the cultures of their peers, teachers and third-country cultures where additional opportunities might exist.

At Sydney we have and continue to give considerable thought to how to provide an excellent learning environment and student experience for both our international and domestic students. A holistic approach is imperative; small-scale initiatives focused on discreet aspects of the international student experience are inadequate. We cannot merely run a good orientation week, provide additional language resources for students from non-English speaking backgrounds, or have international student liaison officers in the student services team, although of course we do all these things. We have developed an approach that supports our international students as an intrinsic part of the institution.

We are focused on giving each student meaningful engagement with the University and our community, both while they are a student and after graduation. This thinking flows across three interconnected themes: an innovative in-class curriculum that better prepares our graduates for future work; a reconsideration of out-of-classroom activity to welcome students and engage with their ongoing social well-being; and a long-term support structure to assist post-graduation.

Within the classroom

We began with a review of an essential part of the student experience: what and how students learn in the classroom and from the curriculum. We identified six critical qualities that our graduates will need in order to be effective global citizens and leaders. These "graduate qualities" are:

Graduate quality	Purpose
Depth of disciplinary expertise	To excel in applying and continuing to develop disciplinary expertise
Broader skills: critical thinking and problem solving communication (oral and written) information/digital literacy inventiveness	To increase the impact of expertise, and to learn and respond effectively and creatively to novel problems
Cultural competence	To work productively, collaboratively and openly in diverse groups and across cultural boundaries
Interdisciplinary effectiveness	To work effectively in interdisciplinary (including inter-professional) settings, and to build broader perspective, innovative vision, and more contextualized and systemic forms of understanding
An integrated professional, ethical and personal identity	To build integrity, confidence and personal resilience, and the capacity to manage challenges and uncertainty
Influence	To be effective in exercising professional and social responsibility and making a positive contribution to society

Source: University of Sydney, Developing a distinctive undergraduate education, Strategic Planning for 2016-20, Discussion Paper No.1, p.10 June 2015

Embedding cultural competence as one of these six graduate qualities is particularly relevant to this discussion of internationalization. As well as being a core competence for graduates, it is an essential part of learning at a university with classrooms as diverse as ours. Every field of study at the University has developed or is revising its curriculum to embed cultural competence and its assessment in its programs. In our Business School, for example, where there is both a high concentration of Chinese students and a great interest among all students in business opportunities in China's rapidly growing economy, there are plans to establish a "China Quotient" program in late 2019. This will deepen students' experience in working in diverse global teams with Chinese peers and include rewards to recognize a familiarity with Chinese individuals, businesses and work contexts.

All students, undergraduate and postgraduate, and our staff now have access to a suite of what we call Open Learning Environment (OLE) units of study. These offer students the opportunity to build novel skill combinations and extend their knowledge by exploring other fields of study. Most OLE units are short, modular courses that are supported by online resources and learning activities and allow students to acquire, in flexible ways tailored to their specific learning needs, foundational concepts and methods of other disciplines. One specific unit is dedicated to cultural competence, encouraging participants to learn about their own identities and how that relates to the wider world. It aims to serve as a starting point in the interpersonal journey of staff and students towards being respectful of diversity and encouraging open, inclusive and interactive behaviour.

In addition to this refocus of the curriculum, we have developed a major commitment to experiential learning both at home and abroad, in real-world settings including the natural environment, community organizations, government or business or within industry. Collectively labelled Industry and Community Project Units (ICPUs), these opportunities place a mix of Australian and international students in multidisciplinary teams and ask them to address real-world challenges identified by industry, government and community organization partners in Australia and around the world. The problems upon which these students work are real strategic problems that are identified by the organizations in Australia, China, Asia more generally, Europe and the US. Their task is to find a solution to those problems, demonstrating the relevance of their own disciplines and a capacity to work in a multidisciplinary team. In their deliberate design, we sought to provide all our students with an authentic international education experience in its own right.

We are also exploring different learning pedagogies through a culturally competent and inclusive lens, critically examining the deeply enculturated nature of our existing pedagogy and considering how best to enable students from diverse backgrounds and with diverse learning styles to develop our graduate qualities. A major current project, for example, is looking at interactive learning experiences and the forms of classroom engagement that best facilitate the active participation of students from learning cultures in which skills such as the ability to speak up and to challenge the teacher are less highly developed than they are among our domestic students.

This overhaul of the curriculum and exploration of pedagogy has presented opportunities to further develop the teaching skills our staff. We provide targeted training on how to put in place evidence-based approaches, facilitate interactions in class between domestic and international students and how to lead inclusive teaching more broadly. We have tripled the number of staff engaging with professional development in teaching through award and "micro-credential" professional learning modules, and continuously highlight inclusive teaching as a core skill for our teachers.

Outside the classroom

Our next challenge has been to create an environment for all students that is welcoming and facilitates and encourage engagement across cultures. As students in Australia typically live off-campus (at Sydney, there are 2,100 beds in dedicated student accommodation on or near campus for a population of nearly 70,000 students), creating a vibrant campus community requires sustained effort. We have redesigned our orientation experience to be more inclusive for students from a wide range of backgrounds. For our international students, many of whom previously reported feeling underprepared for life in Australia and participation in the classroom, we have added in-country pre-arrival sessions focussed on settling into life in Sydney. Our Welcome Week offers a suite of activities not strictly aimed specifically at either domestic or international students but designed to allow each student to design their own experience, with a dedicated app to allow them to develop their personalized schedule. We have also begun to translate our existing support literature into the major languages spoken by our students at home.

As students begin to settle into their new routines within the University, we continue to provide support to smooth transitions and community engagement. Our underlying objectives are to promote greater individual confidence through student social connections, and to support students to develop diverse networks without conscious effort. We have invested in our student organizations to ensure our clubs and societies and their activity programs are accessible and attractive to a diversity of students. In particular, we have promoted discipline-connected societies, so that students with similar academic interests interact with additional co-curricular benefits. We encourage these student groups to run events which focus on industry, civic engagement and career development. We also run a number of peer-led and peer-mentoring programs where senior students support their peers in developing language skills, social confidence and confidence with course materials.

Our initiatives are not focused solely on supporting students through their studies but adopt a holistic approach to student life. We had a wonderful response to a competition with an alumni-funded prize that invited domestic and international students to team up in groups of two to four to create a short (2-minute) film showing life and people in Sydney with the themes of cultures, connections and student life. Another alumni-supported initiative provided small grants (A\$250-\$2,500) to student-led projects to aid cross-cultural interaction. Students have used the funds for projects ranging from trips to Sydney's famous zoo and an outdoor "sculpture by the sea" exhibition, to designing and painting a mural on a new building on campus, to bringing therapy dogs to campus as part of a promotion for mental health support services. Our staff provide training and practical assistance for successful applicants, so that the students running these programs also benefit in terms of the development of their organizational and project management skills. We have also offered free swimming lessons, social sports, and are expanding the opportunities for musicians to collaborate on campus.

Beyond study

Finally, at my University we take enormous pride from the fact that we are educators for our region and our world, and that students are attracted to Sydney from around the globe. Our global network of alumni and advocates is extraordinary and allows us to maintain a global brand that a university in a Southern Hemisphere country of 25 million people would not otherwise be able to do. As a result, we do our utmost to support our students through graduation and once they become alumni in Australia or overseas.

By our best estimate, Sydney has approximately 70,000 alumni living overseas, with the largest concentration (as many as 50,000) in mainland China. As with most universities, we regularly contact them through catch-up events, publications, social media and alumni organizations (including Alumni Groups, and an Alumni Volunteer network). In encouraging these follow-up networks and opportunities, both professional and personal, we find great value in ensuring that the internationalized, culturally competent mindset remains firmly in the minds of our extended University of Sydney community.

CONCLUSION

While all research-intensive universities need to maintain a delicate balance between complex and often contradictory local and international priorities, the situation of universities in Australia is especially precarious. Perched between competing global superpowers in a domestic context that has bound our capacity to fund research inextricably to our ability to attract high numbers of international students, universities in Australia, more than perhaps anywhere in the world, are vulnerable to shifts in the global economic and geopolitical order. As countries around the world, such as Canada and the UK, increasingly look to international student fees as a source of revenue, the situation of Australian universities serves as a sobering example of the inherent risks and limitations of this model.

While most international discussion to date has considered the acute financial risks arising from the "double concentration" issue, the Australian experience has shown that the risks to educational quality and student experience are no less significant. At the University of Sydney, we have thought deeply and invested heavily in developing a unique combination of curriculum, pedagogies and social supports that will meet the needs of an internationalised student cohort, and provide our graduates with the skills, knowledge and values they need to thrive and lead in a rapidly changing world. If we get this right, we may genuinely know what it means to be both a local and a global university, one training our domestic students for life on the international stage, but equally training the future leaders of our region and beyond in an environment that best supports their learning outcomes. It turns out that a situation in which the Australian university system has landed almost by happenstance presents one of the most exciting contexts for higher education innovation in the world today.

REFERENCES

McBride, Edward. (2018). "Australia's economy is still booming, but politics is a cause for concern", *The Economist*, 25 Oct. 2018.

CHAPTER 5

The Global University in the Asian Century

Nicholas B. Dirks

GLOBALIZATION AND HIGHER EDUCATION

In American higher education today, as indeed in American political discourse, there is a palpable and widespread reaction to globalization. A recent front-page story in the *Chronicle of Higher Education* proclaims the end of the global era for education (Chronicle, 2019). Noting a recent drop in the opening of international branch campuses, especially since the presidential election of 2016, it suggests that current political concerns will further depress the international extension and engagement of American higher education. Well before the election, however, there had been signs of a retrenchment. Rick Levin, President of Yale, struggled with significant faculty pushback against his plan for Yale NUS before he stepped down in 2013. John Sexton, President of NYU, encountered mounting faculty discontent in part because of his aggressive pursuit of a global agenda at about the same time. But, every year since, responses to globalization (both in the US and elsewhere) have only intensified, from nationalist and populist on the one side, to solely economic in relation to the spiraling accumulation of wealth by global elites at the expense of the vast majority of the population, on the other. Philip G. Altbach, founding director of the Center for International Higher Education at Boston College, was quoted in the *Chronicle* article as saying that: "The landscape is changing. The era of internationalization might be over, or on life support." (Johnson, 2019)

Meanwhile, the first international programs to disappear *en masse* in US universities faced with budget cuts were related to language instruction; in the last five years, 650 language programs have been discontinued across higher education. This retrenchment has been taking place in a context where only 20% of the population of the US has any familiarity with a second language (compare to Europe, where two-thirds of the population knows more than one language), and levels of bilingual fluency are significantly below other areas of the world where English is not the first language. So although the US has never been very good at promoting the sustained acquisition of "foreign" languages (beyond immigrant groups that nevertheless tend to lose the knowledge of "heritage" languages within an average of three generations), it will doubtless see additional erosion of second language skills, justified in part by the assumption that, in an age of global English and Google Translate, resources for serious language instruction would best be moved elsewhere.

Before World War II, US universities had very little in the way of deep expertise about areas of the world outside Europe and North America. After World War II, however, as the US was thrust into a position of global military and political dominance, the US government allocated significant resources, alongside major investments from some of the most significant foundations (Ford, Carnegie and Rockefeller in particular), to develop a global knowledge base, including in the first instance area studies programs in leading research universities. These programs and centres were designed both to sponsor serious global research in fields ranging from anthropology and history to developmental economics and political science, and to train graduate students in the languages and ways of regions and new nations — many newly established in the wake of European decolonization. The idea was that these students would go on to do much of this research but also teach in colleges across the country, and in this respect area studies were wildly successful. Government initiatives such as the Fulbright program used universities as circulatory nodes for increased global engagement with the goal of building cultural and political understanding, engagement and collaboration — explicitly positioning the United States as the destination of choice for college and university education.

The international recruitment of students and faculty has been a source of great talent creation, not just for the US but for the world; American colleges and universities have not only created more global goodwill but also more economic and social mobility than any other cultural institution or initiative. Millions of citizens from outside the US have been educated in these institutions. Many have stayed and a significant number have contributed massively to the innovation economy of the US, as for example in the Silicon Valley where fully a third of the successful start-ups in the technology world have been led by immigrants.

As universities recognized the extent to which their global recruitment, study abroad and exchange programs, and research relationships served their larger institutional interests (and, often, finances as well), they increasingly sought to take advantage of global opportunities and to expand their "global footprint". In the late 20th century and well into the 21st, US universities began to establish closer partnerships with universities elsewhere, setting up joint programs and sometimes even joint degrees. They also began to build "branch" campuses, sometimes free standing and other times in partnership with global universities with whom they had already established relationships (when it wasn't a purely formal licensing requirement). Qatar's Education City attracted Cornell to build a medical school and universities such as Carnegie Mellon, Northwestern and Georgetown to set up local campuses as well. The most successful branch campuses were arguably set up in Abu Dhabi and Shanghai by NYU, by Duke in Kunshan, China, and in Singapore by Yale in collaboration with the National University of Singapore. Other universities deliberately decided not to build full branch campuses but to set up global centres, allowing minimal investment and maximum flexibility, while also affording opportunities for students, faculty and alumni through the networks these regional centres established and cultivated, including perhaps most successfully Columbia, Chicago and Harvard (the Business School).

As I described in a paper presented at the Glion Colloquium in 2015 (Dirks & Gilman, 2015), I launched an effort some years ago at the University of California, Berkeley, to build what I called the Berkeley Global Campus. The idea was to use a large unused parcel of land belonging to the university on the San Francisco Bay to build a global campus with full participation from top world universities, including Cambridge, the National University of Singapore, and Tsinghua University. We drew up plans for joint research collaborations in areas ranging from global governance and ethics to precision medicine, artificial intelligence, data science, robotics, smart cities, new clean energy sources, climate science and entrepreneurship. The idea was driven by the recognized need to limit overseas investment and political risk, to protect against the possibility of allowing Berkeley's academic and research mission to be compromised by local laws and censorship, and to direct the benefits of global collaboration and partnership to the host campus and the region of northern California, as befits the mission of a public land grant university. Unfortunately, well after the publication of the Glion volume, the plan came up against a continuing financial crisis that hit Berkeley, the preeminent American public university, especially hard. But it was also affected by a reaction to my plan to expand the global mandate of the university.

Clark Kerr, the first Chancellor of Berkeley, noted some 50 years ago that "the university is so many things to so many different people that it must, of necessity, be partially at war with itself." This war continues to rage, as many universities struggle to define what the 21st century "multiversity" (to use Clark Kerr's famous nomenclature) needs to be, one that will be not just more networked and permeable, but also more global. While I understand and share the critiques of globalization that focus on growing inequality and massive disparities of wealth creation, not to mention the extent to which some university ventures to create branch campuses have run into major financial and political difficulty, I find the impulse to jettison efforts to enhance our global connections to be retrograde at best. All of our major challenges are now global challenges, and, whether we like it or not, the only meaningful solutions to the problems we face will be global in form and substance — and this includes the educational challenges ahead of us all. I believe that a genuinely global strategy for educational institutions is both inescapable, and a necessary component of any effort to reimagine the future, not just for colleges and universities but even for education at earlier levels as well.

CHINA AND AMERICAN HIGHER EDUCATION

There are multiple reasons to embrace a global agenda, and not only because over the last several decades American (and British, as well as Australian) universities have become dependent on the billions of tuition dollars (US$39 billion in 2017) coming from what until just a year ago was a steadily growing international population of students (in 2017-2018 there were over a million international students studying in the US). However, it is worth noting in this context — to focus now on China, which produces the largest number of international students in the US — that while the current "trade war" with China has been seen as predominantly about cars, soybeans, steel and technology, less evident but no less important is the critical role of education. In 2017-18, slightly more than 360,000 students from China were studying in the US, one third of the total number of international students (the next closest number is the 200,000 students from India). One third of students from China are undergraduates, close to a half are doing post-graduate work either for Master's or PhDs, and one sixth are in K-12 schools, mostly secondary. Chinese students have favoured the US as a destination for college for some time, and there are reasons to worry this may be adversely affected by the present conflict. Meanwhile, the numbers of students from India has risen in recent years as the US has slowly displaced the UK as the destination of choice, while undergraduate options in India were challenged by the decline in many traditional institutions of higher education, but in the short term China plays the largest single role in international student numbers. And it is not just the tuition dollars that contribute to university life; many scientific labs in our major research universities would stop functioning without Chinese graduate students and post-doctoral fellows.

The financial contribution, however, is extremely important, especially at a time of stressed university budgets, particularly in public research universities. Given this background, it comes as little surprise that the schools of business and engineering at the University of Illinois started paying $424,000 last year for an annual insurance policy against a possible decline of Chinese students to protect against losses up to $60 million. Given the numbers of Chinese students at leading research universities, any significant drop in enrolment of these students could be devastating for precarious university budgets, whether public or private. And, although India is not currently targeted in the US in the same way as China, the rise of political concern about international students could spread beyond China at any point due to changing global economic or geo-political conditions.

Although the persistent rumour that the Trump administration was considering a ban on student visas for China was quickly quashed, such a ban was apparently under genuine consideration from some in the White House, and reflects a continuing concern about the possibility of espionage and property theft occasioned by the large number of students from China engaged in advanced graduate training and research in fields ranging from computer science and artificial intelligence to biotechnology. The FBI has lately been cancelling the visas of an increasing number of Chinese scholars, including social scientists with deep knowledge and appreciation for the US. And, more recently, government officials have been visiting research universities, warning them of potential dangers and suggesting increased monitoring of students from abroad. Should students ever become embargoed as part either of national security concerns or the trade war broadly conceived, it is highly unlikely that the Trump administration would provide additional subsidies to offset these losses, in the manner it is doing for farmers in the Midwest affected by soybean tariffs.

For decades, Chinese students have sought admission to US universities because these universities have been the gold standard for both education and research. Ever since there were global rankings, US universities have dominated the world stage, attracting not just students but world-class scholars and researchers from around the world. But the Chinese state has been investing heavily in its universities, and top Chinese universities have not only climbed in global rankings but successfully begun to recruit leading scientists back from US and UK universities. Last year for the first time, Tsinghua University in Beijing ranked number one among all Asian universities. Tsinghua took the top position from the National University of Singapore, that had led Asian universities for several years after toppling Tokyo University. But Tsinghua's rise has not happened only because of major state investment in faculty and facilities, but also, I would argue, because of a systematic strategy of global engagement (as both NUS and

Tokyo University had done before). I came to appreciate this objective as Tsinghua partnered with UC Berkeley in building the Tsinghua-Berkeley-Shenzhen Institute in 2015, while also establishing joint programs with other world-class universities such as the University of Washington and building the Schwarzman College to bring outstanding young college graduates to Beijing for a year of study.

When Chinese students travel to the US for their education, and when Chinese universities pursue global engagement, they do so because they are seeking the best opportunities for education and for research. They also follow a pattern that other countries have used at similar stages of their history. I have already rehearsed the modern history of global engagement for American universities, but it is important to recall that these same universities grew from small and largely provincial undergraduate colleges in the 19th century to become major world-class research universities in large part because of the influence of German universities, which many leading American educators and scholars attended in the late 19th century. Whatever the form of global influence or engagement, the most successful universities both in educational and research terms are those that have been open to new ideas and human capital coming from all over the world.

The dramatic increase in the quality of Chinese universities may by itself lead to a time when fewer students travel abroad from China for their education. When combined with the escalation of rhetoric around the trade war with China, however — especially the recent attribution of espionage and intellectual property theft associated with Chinese students and scholars on American campuses — this trend could become increasingly precipitous. But my real point here is that the US has more to fear than a slow diminution of tuition dollars coming from China. The value of these exchanges is far greater than monetary alone; first and foremost, the academic and research value has been and continues to be enormous, as universities recruit top talent from global pools of candidates. Additionally, the friendships and networks established during study abroad can last a lifetime, resulting in political alliances, business relationships and further research collaborations. The relationships established with universities on a global basis can also result in major philanthropic contributions to *alma mater*, which is another reason why university leaders from the US frequently spend so much time travelling in Asia. And it is clear that the Chinese scholars whose visas are being cancelled have contributed not just understanding but appreciation for the US, whether for its universities or for its society, culture, and (at least until recently) its political system.

Trade too has benefits that go well beyond the immediate economic returns of trade. Long ago Adam Smith, canonic champion of free trade, recognized that trade produced "sympathy" — by which he meant cultural recognition and understanding across distant populations. Whatever the truth of that assertion, this in fact applies far more so to the world of education than to any other domain of human exchange. Given the tensions between the US and China at the current moment, the relationships that develop because of student and faculty interactions, and as a consequence of educational as well as research collaborations, are especially valuable. Leaving aside the dangerous possibility of increased conflict — whether economic, political, or military — almost all of our major challenges now are global challenges, requiring global solutions. There are multiple reasons that the current escalation of suspicion, and single-minded focus on the security risks of educational exchanges and collaborations, is short-sighted at best.

UNIVERSITIES AND GLOBAL CHALLENGES

It goes without saying that the more understanding we have in this world, the better off we will be. Educational exchanges, collaborations and networks increase international understanding as well as creating life-changing personal relationships and interests. But it is important to stress that this is true in the domains of research and public service as well as education. If, to take perhaps the most obvious examples, we are to begin to tackle climate change, or global public health challenges, or even global inequality and some of its most direct effects, we know we need to do so across national borders if we are to be effective, for no wall or barrier will keep a global pandemic or carbon dioxide or a rise in sea levels from being global migrants. And for all of these challenges, educational institutions can be primary ambassadors of global cultural understanding and cooperation.

We have stressed the advantages of global approaches for university budgets and advanced research, but it is important as well to acknowledge that the global circulation of students, faculty and ideas about teaching and will ultimately be necessary for our educational institutions themselves to adapt and to thrive in a global marketplace, even in the face of new funding challenges and increasing demands for accessibility and affordability. New ways of thinking about student achievement and learning, about the relationship of cognitive and behavioural development and the best strategies for teaching, about the ways in which certain kinds of applied or vocational skills need to be supplemented with softer skills in order to translate into a lifetime of meaningful employment and constructive societal contribution, about how to transmit creativity and imagination, about modes of assessment that can be adapted to localities but also translated in global contexts, about how better to align the methods and strategies of education at all levels, among many other things, will be more productive if engaged in ways that bring together global resources, ideas and institutions. In short, to find and adopt best practices requires being able to extend one's reach across the globe.

In addition, the growing insularity of campus culture in many regional contexts can only benefit from more rather than less global interaction. Students
who have used their college years to cultivate local forms of identity politics
are often unaware of the limited provenance of their own political concerns
and debates. New forms of solidarity, along with renewed recognition of the
wide range of cultural difference that exposure to the world introduces, can
only expand the horizons of new generations of students. And while it was
understandable in the past that concerns about issues ranging from academic
freedom to freedom of speech and even critical thinking have typically been
mobilized against interactions with universities in places such as Singapore
and China, the sad truth is that all of these concerns are now universal and
will only benefit from more global exchange to deal with the growing challenges of resurgent ethno-nationalism, political populism, the widespread
return of authoritarian models of governance and the pervasive (and not
unrelated) effects of social media on political life.

THE ASIAN CENTURY

There is, however, another reason why colleges and universities in the US (and
other parts of the West) need to resist the call to national retreat, even when
muted in the politically progressive tones of places like UC Berkeley which
understandably has a primary obligation to state level constituents. And this is
the fact that we are in the beginning phase of a transition from the American
Century (christened as such by Henry Luce in an influential article in *Life*
magazine in 1941) to what is now arguably the Asian Century. This is already
reflected in the level and scale of resources being mobilized to support higher
education, especially in China, but increasingly in other Asian countries and
centres as well. It is also determined by major economic and demographic
trends. The number of new cities in China with over 10 million people has for
some time far eclipsed the number of old cities in the US with similar populations. But, in the next few decades, India's population will overtake China and,
from all reports, its economy will continue to grow quickly as well, not least
because of the burgeoning middle class. This demographic transition — when
coupled with the rapid creation of new wealth and the high value placed on
education — will have untold effects on the world of higher education.

At the same time, we are already seeing the development of an enrolment crisis in a growing number of liberal arts colleges in the US, especially
in the Northeast and mid-Atlantic. While this crisis has been generated in
part by a growing concern about the cost of higher education in the US,
along with similar concerns about the economic returns of traditional liberal
arts degrees, it is also about demographic shifts in the US to southern and
western states. And it is likely that this crisis will expand to large research

universities as well if there is a real drop in the numbers of students from Asia coming to the US for study, for undergraduate as well as graduate education.

Even as the rise in educational and research quality in China and other parts of Asia is a necessary element in the actual realization of a new century that might be dubbed "Asian" for its dominant forces and influences, we also know that future trends will make Africa ever more important, in the first instance because of rapid population increases and the growth of middle class markets and lifestyles, but also for reasons having to do with the possibility that the kind of stagnation that places like Japan have experienced might spread across other parts of Asia, North America and Europe. That being said, climate change is likely to displace not only larger and larger populations but to create other disruptive geo-political trajectories as well that will have unpredictable consequences for the global balance of power.

THE GLOBAL MULTIVERSITY

Instead of responding to the current moment by retreating from educational globalization, therefore, I would propose that we imagine a different level of global engagement altogether. I have recently been involved in advising a new educational effort to build a global network of K-12 schools extending across Asia, the Middle East, Europe, the US, Africa and Latin America. As a result of this experience, I have been wondering if a new kind of global multiversity — a genuine network across regional/national boundaries and borders — might be possible. If, as in the school project I've been working on, one could in fact design a single university with multiple campuses in different countries, one could think quite differently about disciplines, academic structures, the nature of foundational knowledge and the relationship between the development of knowledge expertise and readiness for the world after college. One could engage in the fantasy of many a university president — the building of a new university from scratch. One would of course do so with constraints and guidelines predicated on the examples set by the world's great universities, though one would also wish to draw from models that come from some of the most dynamic examples of universities that have significantly improved their standing because of their capacity to change dramatically under dynamic leadership (e. g. Arizona State University and Northeastern University in the US).

As we are about to enter the third decade of the 21st century, however, any new university should be less dependent on national models than on the recognition that successful universities for some time — and certainly in the future — must and will be global in that they must perforce appeal to a global set of constituencies, including multiple bodies, from governments and regulatory bodies to corporations and potential industrial as well as non-governmental partners.

There is also no doubt that any new university — and this of course simply reflects the demographic and economic realities alluded to above — will have to draw on a global population of students and faculty if it is to have the capacity to thrive in the coming century. It must have the capacity to draw on global resources to support the kind of research that will be necessary to maintain research relevance and excellence at a level and on a scale to compete with a growing number of excellent educational and research institutions across the world.

The purpose of this article is not to propose a design for a global multi-versity, but rather to suggest that any new models for higher education need, among other things, to ensure that the institutions we build for the 21st century and beyond take on the global in a more concerted, systematic, and even ambitious way that we have in the past. And, although the difficulties of changing and adapting well-established institutions are keenly appreciated by all of us attending the Glion Colloquium, the real point here is to suggest the importance of maintaining and expanding the global footprint and connectivity of all of the institutions we lead, wherever we might be located, and whatever level of reaction to the global dimensions of the fundamental mission of knowledge acquisition and dissemination might be directed towards the university.

REFERENCES

Chronicle. (2019). "The end of a global era: How grand plans for international education are fading", *The Chronicle of Higher Education*, 5 April 2019, Volume 65, Issue 29. Online: https://www.chronicle.com/issue/2019/04-05

Dirks, N. & Gilman, N. (2015). "The Evolution of globalized Higher Education", In *University Priorities and Constraints*, Weber, L. E. & Duderstadt, J. J., Eds, Economica, Paris, London, Geneva.

Johnson, S. (2019). "Visa Woes, Politics, and Fears of Violence Are Keeping International Students Away, Report Warns", in *Chronicle of Higher Education*, 29 May 2019. Online: https://www.chronicle.com/article/Visa-Woes-PoliticsFears/246398

CHAPTER 6

The role of a rising university in an emerging international metropolis

Shiyi Chen

THE UNIVERSITY AND THE CITY IN THE AGE OF KNOWLEDGE SOCIETY

Colleges and universities are among the oldest type of organization, originating from the medieval age. In the long historical trajectory of this social institution, it is commonly recognized that colleges and universities evolved through three stages — from the British model of gentlemen education, through the German model of scientific research, to the American model of social service. A research university is nowadays a combination of the three. Most distinctively, the history of this evolution is also an irreversible path on which the university was transformed from an ivory tower to a social institution critical to social-economic development and the welfare of the human race.

The Glion Colloquium, launched in 1998, has closely captured the evolution of research universities in the past two decades. As succinctly summarized by Peter Scott in his 2015 review of the Glion Colloquium contributions, the 21st century is characterized by a global "knowledge economy" and "knowledge society" in which universities partner up with industries, actively engage in communities and have taken up a central position in a society, economy and culture shaped by globalization and global competitiveness (Scott, 2015, pp. 42-44). Against this backdrop, many Glion colloquium participants have noticed the rise of Asian universities, propelled by heavy investment and unreserved support from governments that are determined to achieve economic growth and competitiveness through innovation stemming from university research. Howard Newby's and Peter Scott's contributions in 2015 both pointed out the contrast between the dwindling public funding and the intensified public doubt about universities in the West and the strong government support and public valuation of universities in Asian countries with universities on the rise. Newby even titled his 2015 contribution "The Divergent Fortunes of USA, Europe and Asia" (Newby, 2015, p. 53).

THE NUANCES OF THE CHINA STORY

While governments and the general public in the West may indeed learn from Asia, it has historically been instrumental for Asian universities to learn from the universities in the Western world, especially those in the US, in order to realize their ambition to become "world-class universities". In China, the tradition for Chinese students to pursue their advanced study in Western universities and of Chinese universities to be staffed by overseas returnees who were educated or trained in Western universities has lasted for a century. Since the beginning of the new millennium, the Chinese central government has been heavily investing in the development of public universities governed by the Ministry of Education in order to avoid brain drain and to develop the capacity to cultivate local talents for national goals. Subsequently, the central government launched a series of initiatives in pursuit of "world-class universities" (Lin, 2017, p. 30). The government commitment has included not only funding but also policy support. The Chinese government has learned from extensive studies of the world's best universities that the advancement of universities cannot be achieved by relying on monetary investment solely, but must take into consideration the institutional structure and work culture, including governance, management, academic norms and professional ethics. This new approach has been captured in the Chinese government discourse by the term "the modern university system" (Lin, 2017). The central government's ensuing encouragement of system reforms and innovations inside and outside universities is inseparable from the legitimacy provided by the world's best universities in the global arena, and enforced through the universities' international exchanges and collaborations.

Dr Bernd Huber, President of Ludwig-Maximilians-Universität Munich, defines "the model of the modern university" as including essentially institutional autonomy, academic freedom, peer review and the embracement of competition (Huber, 2015, pp. 69-70). Both Scott and Newby expressed doubts about how much China has implemented "the mode of the modern universities" through the isomorphism process (DiMaggio & Powell, 1991), and, very noticeably, both point out that the key concern is about some normative issues, such as the adoption of academic freedom, gender equity, etc. However, with 20 years' experience working in the United States and 14 years of working in China, I would like to affirm that if one looks into the complex reality of the innovative Chinese universities, one will see clearly that their learning from Western universities has delved into the normative depth of the academic profession. A crucial factor that enables this normative isomorphism is globalization, which effectively removes the walls between countries not only for mobility, but more importantly, for sharing values and ideas, and for the formation of a global academic profession.

In Scott's review, he lists the mobility of students and academic staff, the establishment of offshore campuses, and world university rankings (Scott, 2015, pp. 33-34) as prominent features of the globalization of higher education. While, according to Newby, globalization is in general perceived as "pre-eminently an economic and technological phenomenon" (p. 43). To the rising Chinese universities, globalization can mean much more than that. I would like to share with the reader my observation of several new rising universities in China and, in particular, my own experience of building a new university, the Southern University of Science and Technology, in an emerging metropolis, the City of Shenzhen. I hope my contribution can enrich the analysis of the evolution of universities and contextualize the phenomenon of the rise of Chinese universities.

THE NEW INNOVATIVE UNIVERSITIES IN CHINA

Thomas Bender, the renowned historian at New York University, argues that a city without a major university is an incomplete city (Bender, 1991). Along with the economic development of Chinese cities, and cognizant of the huge gap between Chinese universities and the world's best universities, several major cities in China started to build whole new local universities with an ambition to develop them into institutions of world-class quality in a short span of time. Examples are SUSTech in Shenzhen in 2011, the University of the Chinese Academy of Sciences (UCAS) in Beijing in 2012, Shanghai University of Science and Technology (ShanghaiTech) in Shanghai in 2013 and West Lake University in Hangzhou in 2018. All four universities are supported by their local governments. SUSTech is 100% funded by governmental appropriation. UCAS and ShanghaiTech rely on the Chinese Academy of Sciences system to quickly assemble the necessary factors for operation, such as the faculty team, lab facilities, and degree conferral qualification, but they both receive capital funding from the local government to be able to build a state-of-the-art new campus, and to offer a competitive compensation package to overseas returnees. These three are public universities. In contrast, West Lake is a private university starting initially only with gradu-ate degree programs; undergraduate education might be added several years from now. In this case, it is again the local government that provides land and initial funding for buildings and research.

Another observation relates to the urban context. The city of Shenzhen is known for its paucity of higher education institutions in comparison to its large and young population and the prosperous economy dominated by successful private corporations. But the other three cities have already had world-renowned universities, such as Peking University and Tsinghua University in Beijing, Fudan University and Shanghai Jiaotong University in Shanghai, and Zhejiang University in Hangzhou. So why did they still commit to the creation of one more university? The answer is to create innovative universities that can truly adopt and implement a modern uni-versity system to ensure their success. The established universities carry too many conventions and history to reform their internal structures or really adopt practice proved to be successful by the world's top universities, such as the tenure-track system, the PI system, faculty governance of academic affairs, capability of generating high quality publications in the top English-language journals in the field, and a low student faculty ratio for substantial student faculty interaction, to name just a few.

These new universities almost exclusively focus on science and engineer-ing. Not only are these subjects most directly pertinent to economic devel-opment. There have also been several successful precedents in the world which proved that the ambition of quick and major achievement is feasible. The role models include universities such as Warwick University in UK, Nanyang Technology University in Singapore, Postech in South Korea and Hong Kong University of Science and Technology in Hong Kong. Especially the success of the latter three Asian universities demonstrated to the spon-soring government and the founders of the new universities the effectiveness of borrowing from the Western university system.

While the central government has increased funding and policy com-mitment since 2000 to those universities under its governance, it has paid more attention to policy compliance and quantitative evaluation. The role the central government plays is that of a regulator. A significant difference between the central and local governments is the fact that the latter have a stronger sense of ownership in the higher education experiments and are more likely to form a real partnership with the universities. The local gov-ernments care more about the practical output and real impact of these new universities. On one hand, the universities work very hard and spend one year like three years; on the other hand, they attach great importance to media strategies in order to gain more confidence from the governments by generating positive publicity. By referring to the successful universities in the world, they also work on persuading governments to be more patient and remind the latter of the importance of arm's length. As suggested by colleagues at the Colloquium, the "partnership" and the universities' institutional autonomy will be tested when the governments try to hold the universities accountable through any measurements or when the universities develop beyond the interests of the local governments. There is still a long way to go to generate true partnership between these universities and their local government, or to pave the policy ground for institutional autonomy, which shall be the very foundation for a sustainable development of these universities in the next 50 years toward excellence.

The above context is crucial to an understanding of the current advancement of the new Chinese universities. Very visibly, they all feature lavish government investment, but more significantly they are also sustained by institutional innovations, internationalization, and a close partnership with their cities. These features are most distinctive in the case of my own university, SUSTech, located in the City of Shenzhen, which I would like to use as an in-depth case study. In comparison with the other three peers, it has a simpler governance relationship with the government, plays a more instrumental role for the future development of the city and resonates more with the trailblazer spirit of the city itself: Shenzhen was established as the first special economic zone in China, the first window to the world in the post-Mao era, and the cradle for the Open and Reform Policy of China.

SUSTECH AND THE CITY OF SHENZHEN

The idea of establishing SUSTech as Shenzhen's first research university was formed in 2007 by the municipal government when the city was officially 27 years old as China's first special economic zone and 28 years old as an administrative division on the Chinese city map. It had started in 1979 with a population of 300,000 and a rural economy featuring fishing. In 2007, the City had still only one teaching university, one polytechnic college for associate degrees, and three graduate schools as the branch campuses of Chinese universities from other cities. Since the higher education system was not able to catch up with the economic and urban development of the City, the City had favored a "borrowlism" strategy by inviting famous universities in other major cities to establish their branch campus in Shenzhen, which was a fast approach to address the needs of fundamental research and high-level talents. In 2007, the municipal government eventually determined to invest in the creation of a local research university that could provide the original knowledge, technology innovation and talent development for a sustainable future of the city.

In 2009, SUSTech appointed its first president. In 2011, located on a borrowed campus, it had its first cohort of 44 undergraduate students, a dozen faculty members, four departments, five degree programs, and a budget of US$15 million. In 2012, it was officially recognized by the Ministry of Education. In 2012, the city had a population of 14 million with an average age of 27; the 4th largest GDP, and the most vibrant market economy with the largest number of private companies among Chinese cities.

In 2017, Shenzhen set the goal of becoming an international innovation hub for high-tech industries. In 2018, the City declared itself an international innovation hub for science and technology. Interestingly, the same year, Shenzhen was ranked No.2 by *Lonely Planet* among the ten cities in the world that are most deserving to be visited. The city now has a population of 21 million with an average age of 33; the 3rd largest GDP, the most fully developed industry supply chain among Chinese cities, and was ranked the 14th financial centre globally by the Global Financial Centers Index (London), released in March 2019. In particular, Shenzhen has been known as China's Silicon Valley but with strengths in both hardware and software, and has been the cradle to multinational corporate giants such as Tencent, Huawei and DJI.

In fall 2019, eight years after the enrolment of the first class, SUSTech has 4,205 undergraduate students, 2,214 graduate students (majority PhD students), 800 faculty members (about half are tenure line faculty members, half research and teaching faculty members), 15 departments and 29 degree programs that cover sciences, engineering, business, life science and medicine, with a budget close to $500 million and a campus with construction areas of 522,000 square metres (to be doubled by the end of 2020).

The City of Shenzhen is gradually ascending to the status of an international metropolis. On 18 August 2019, the State Council of China issued a monumental directive to designate the City of Shenzhen as an exemplar city in China to pioneer and showcase the development of advanced urban civilization. The 30 areas in which this directive eagerly propels Shenzhen to excel include a national scientific research centre, medicine, creative design, financial market, digital currency and mobile pay, innovative digital economy, ecocivilization, talent policy, deep ocean research, to name just a few. Many of these areas, if not all of them, cry out for a prominent role to be played by research universities for the advancement and sustainability of the city, the country and the human society. It is especially encouraging that the directive reiterates the importance of sticking to the course of internationalization and open-door.

The University has benefited enormously from the City's steady growth and rising status. Although SUSTech is still too young to see a large number of distinguished alumni contribute to the development of the City, for three years in a row the University has been acknowledged by the City as the best talent-recruiting institution for the remarkable number of its senior academic hires. The University has made remarkable progress in building strong research programs and state-of-the-art facilities in particular areas, such as the third generation semiconductor, quantum physics, brain research, artificial intelligence, robotics and advanced manufacturing. They will enable the City to venture into the future frontiers of technology innovation, in the next era of economic development of the Greater Bay and of globalization.

Needless to say, the development of the University is also a process to obtain an indigenous adaption to Chinese society by working on problems imperative to the local area. For example, the University's College of Environmental Science and Engineering established the Institute of Research on Sustainable Development to address the urban and environmental problems while Shenzhen is fast growing into a mega-metropolis. The University states its development principle as being "rooted in China and striving to achieve world-class quality".

More comprehensible for a wider public is the rise of SUSTech in the university rankings. According to the 2019 and 2020 World University Ranking by Times Higher Education, SUSTech was ranked No. 8 and No. 9 respectively among mainland Chinese universities, with the highest publication quality in China, and ranked between 300-350 in the world. The THE Young Universities Ranking has SUSTech at No. 55 in the world. *Nature Index 2019* placed SUSTech 28th in China, 183 in the world, 4th in the list of Rising Stars. The international rankings are really helpful to the University to flag up for the general public who we are, and for the government to know where our standing is in the university system. This is especially important since SUSTech is not a "985" or "211" project institution nor does it belong to the "Double First Class University Plan" in the MOE-managed system.

I think it is fair to say that we have seized the historical opportunity China has offered to her higher education. More importantly, we must have done something right. Among Chinese public universities, SUSTech has the only governance system featuring a Board of Regents, a collective board comprised of the University senior management and other representatives, the executive of the Municipal Government, and leaders of the larger social sector, from business and education. We are also the only Chinese public university that selects its own president through the Board of Regents rather than accepting appointment from the government, and the first Chinese public university that admits 100% of the students not solely through the national college entrance exam (Gaokao) but a rigorous admission procedure with all-round assessment (the "631" model) that is based 60% on the Gaokao, 30% on a SUSTech-administered test and interview, and 10% on the previous high school performance.

The University is international, English-speaking, innovative and entrepreneurial. The overall strategic development of the University is guided by an International Advisory Council (IAC) comprised of 16 university presidents or former presidents; 90% of the faculty members have been trained or had previous appointments abroad, 60% of them from the world's top 100 universities, and about 30% are foreign or Hong Kong passport-holders. All tenure line faculty members receive generous start-up funding so they can focus on meaningful work. Teaching affairs, hiring and promotion are decided by academic committees constituted of academic department chairs. Faculty members collectively design the curriculum and individually decide how to teach, while being evaluated by students and faculty peers.

More than 70% of the required undergraduate courses are taught in English. The students enjoy a student faculty ratio of 10:1, a living learning environment that relies on both classroom learning and a residential college system for whole-person development. All faculty members, including the university's senior academic leaders, are assigned to a residential college as faculty advisors; 100% of the undergraduate students participate in research. Study-abroad programs are an essential part of the undergraduate education. The students do not have to claim a major until as late as the end of their second year and are strongly encouraged to look for their true passion, using the help and clues they can get through research, advising or individualized course taking. The science and engineering curriculum is complemented by a large number of course choices in the humanities, social sciences and arts. SUSTech is the first Chinese public university that has implemented a requirement of writing courses to train the students in critical thinking and communication.

Both faculty members and students are encouraged to engage in entrepreneurship. Faculty members are assisted by the Technology Transfer Office to collaborate with industries or start their own spinoffs. Since 2015, the University has seen the establishment of more than 50 companies. Due to the fact that we foster relevance to the local economy, the students have more internship choices and receive a variety of real-world problems contributed by local companies for their senior capstone project.

The university has implemented international models of university governance and education within a Chinese context to ensure institutional autonomy, student-centred education and high-quality research. The international partnerships provide us with the evaluation, recognition and endorsement by the top universities outside China, and, what's most important, legitimacy and protection. SUSTech is able to innovate in ways very different from established Chinese universities, while still being acceptable to the sponsoring local government and being in alignment with the monitoring government agencies at the provincial and national levels. In 2016, China's Chairman, Xi Jinping, depicted a conceptual framework for development which emphasizes the leading role of science and technology, and the propelling power of innovation. This national discourse enforces what has been embedded in the mission of the University, namely to serve the City's sustainable socio-economic development and to be the engine that propels the City's continuous prosperity. Through the success of its actions, SUSTech has built confidence within the Municipal Government in the importance and the promise of universities for a city's prosperity. As a result, Shenzhen now plans to invest in the establishment of several additional new universities over the next five years.

RESEARCH UNIVERSITIES AND THE FUTURE OF THEIR GLOBAL COMMUNITY

We at SUSTech are not worry-free. It was true in the past that the idea of the university embraced a knowledge production and dissemination that were open to all. It is also true now that the relationship between universities and the knowledge society seems to create boundaries for what can be shared publicly and what cannot be. Professor James Duderstadt in his 2017 contribution to the Glion Colloquium succinctly summarized:

In this knowledge economy, where the key assets driving prosperity are intellectual capital, education has become a power political force, both nationally and on a global scale. (Duderstadt, 2017, p. 194)

This is the political landscape in which the research universities of the world are now situated, perhaps partly because we have done too good a job in serving economic society. The innovative new Chinese universities are the intellectual offspring of the modern university, the latter's indigenous adaptation in China and a new member of the international community. These universities maintain the openness of Chinese society to the world. That is perhaps true to the role of any university playing for its country and culture. It will be devastating to the universities in any country to sever the exchange and communication with the global scholarly community.

What the rise of Chinese universities mean is yet to be determined by how much significant contribution they can make to the human world, but not by the numbers of publication or citation index. I hope that the case study of SUSTech and the City of Shenzhen may be of help in our examination of the current situation, as an example of how indispensable research universities are to human society and a sustainable world.

REFERENCES

Bender, T. Ed. (1991). *The university and the city: From medieval origins to the present.* Oxford University Press.

DiMaggio, P. & Powell, W. (1991). "The iron cage revisited: Institutional isomorphism and collective rationality in organization fields". In *The institutionalism of organizational analysis*, DiMaggio, P. & Powell, W., Eds, The University of Chicago Press, Chicago.

Duderstadt, J. J. (2017). "Preparing the American University for 2030". In *The future of the university in a polarizing world*, Weber L. E. & Newby, H., Eds. (Glion Colloquium Series No. 11, pp. 67-81). Geneva.

Huber, B. (2015). "The future of universities: Academic freedom, the autonomy of universities, and competition in academic revisited". In *University priorities and constraints*, Weber, L. E. & Duderstadt, J., Eds. (Glion Colloquium Series No. 9, pp. 67-81). London, Paris, Geneva.

Lin, J. (2017). "The evolution and missions of universities in China". In *The future of the university in a polarizing world*, Weber, L. E. & Newby, H., Eds. (Glion Colloquium Series No. 11, pp. 29-34). Geneva.

Long Finance (2019) GFCI 25 Rank. Archived on https://www.longfinance.net. Retrieved May 2019.

Newby, H. (2015). "Global diversity in higher education systems: The divergent fortunes of USA, Europe and Asia". In *University priorities and constraints*, Weber L. E. & Duderstadt, J., Eds. (Glion Colloquium Series No. 9, pp. 53-66). London, Paris, Geneva.

Scott, P. (2015). "Glion colloquium: A retrospective." In *University priorities and constraints*, Weber, L. E. & Duderstadt, J., Eds. (Glion Colloquium Series No. 9, pp. 23-50). London, Paris, Geneva.

PART II

•••••••••

The Local

CHAPTER 7

Science systems under pressure: The entrepreneurial *must* of traditional universities in the 21st century

Andrea Schenker-Wicki

THE EVOLUTION OF UNIVERSITIES

After the fall of the Roman Empire in the 5th century of the Common Era, the public education system almost completely disappeared, with just a few church schools remaining. It was not until centuries later, when towns and international trade began to flourish again, that the value of education was recognized, and illiteracy addressed across a broad front. For this purpose, the first citizen schools were established. These schools, together with the church schools, subsequently evolved into universities. Students and teachers at these institutions formed a community, a collective, in other words, a *universitas*. In the Western world, the Pope and the Emperor protected these new institutions and granted them special privileges: each university had its own jurisdiction and autonomous governing body, making it almost a state within a state. The universities' main mission at the time was education, to which much importance was attached. After teaching had dominated at universities for almost half a millennium, the understanding of science changed in the 18th century, and experimental research became more important. This type of research led to a sharp increase in the number of professors and consequently to the creation of individual faculties. At the same time, increasingly more students were attending university. Since the end of the Second World War and, in particular, in recent decades, universities have evolved from elite institutions where only a small percentage of the population was educated (approximately 5% of the corresponding age cohort) into a universal system that encompasses 40%-50% of young adults (Trow, 2007).

Eventually, the self-governance and autonomy of universities was brought to an end in the 19th century. The majority of publicly funded universities were integrated into the ministerial bureaucracies of their states, meaning the loss of the autonomous status that they had enjoyed for centuries. It was only in the last decade of the 20th century — and slightly later in Germany and Austria, at the beginning of the 21st century — that universities in continental Europe regained some of their autonomy. The aim was to transform the rigid, bureaucratic systems into efficient, effective and profitable service facilities. Under the title New Public Management (NPM) — or results-based management — new governance tools were introduced, legislation streamlined and modernized, and responsibilities redefined between the executive, legislative and administrative bodies. These reforms were aimed at giving universities more autonomy to stimulate the creation and dissemination of knowledge and innovation. In Switzerland, the term "autonomy dividend", associated with more efficient "knowledge production", was used. It was also assumed that the new-found freedom would give universities more leeway, particularly with respect to financial management, which they could then use to their advantage (Schenker-Wicki & Olivares, 2010). However, the extension of autonomy in the areas of organization and finance — mainly under the headings "performance agreement" and "global budget" — had a downside: the additional autonomy drastically increased the accountability of universities in a number of respects. The institutions concerned were obliged to introduce comprehensive reporting to measure and assess their activities (Haldemann, 1998). The form that accountability took varied greatly within Europe and depended on how much trust was given to the individual university by the responsible government agency.

However, the reforms did not affect the general consensus in continental Europe that education was a public good. Much of this understanding was based on the positive external effects on the economy that result from a competitive stock of human capital ("capacity building") (Weiss, 2000). As a result, education in continental Europe was (and still is) largely subsidized by the public purse. Therefore, tuition fees at universities in continental Europe are relatively low compared with those in the US or the UK. Some German-speaking countries have even abolished tuition fees altogether. Conservative governments in Austria and Germany introduced moderate tuition fees for publicly funded universities, but these fees were quickly scrapped as soon as a social democratic party came to power. Despite the intensive discussions about substantial increases in tuition fees that have taken place repeatedly in German-speaking countries, the paradigm of education as a public good has held firm. At present, it is politically impossible (at least in German-speaking Europe) to propose that students make an increased contribution to cover the costs of universities.

THE NEW CHALLENGES FACING THE ECONOMY AND SOCIETY

However, it is not just the universities that have undergone a drastic change in recent decades — the society and the environment in which they operate have also evolved. The challenges facing research universities in the early 21st century are of concern to the governing bodies of many universities. They essentially relate to three developments: globalization, leading to an unprecedented acceleration in the pace of life; demographic change, associated with an aging society; and the increasing importance of the knowledge society.

Globalization: It is appropriate to begin with globalization. Globalization has drastically increased the speed of many daily and work-related processes, primarily due to the high concentration of different potential interactions. The megatrend of globalization goes hand in hand with a huge acceleration in knowledge generation. Never before has so much new knowledge been created, meaning that what was correct and relevant yesterday is outdated or irrelevant today. In the technical professions, the half-life of specialist knowledge is estimated to be approximately five years, indicating that acquired knowledge loses up to 50% of its relevance after this period, as it is replaced by new findings (Schüppel, 1996).

Demographic change: The second major development is demographic change caused by the decrease in birth rates and the increase in life expectancy, which will pose problems not only for Switzerland, but also for the whole of Europe. In Europe, the employment to pension ratio will shift from 4:1 at present to 2:1 by 2050, and the working population will shrink from today's figure of approximately 310 million to 250 million (Eggenberger, 2015). This forecast also applies to Switzerland, where the number of retirees will increase by more than 50% in all cantons over the next 10 years (Swiss Federal Statistical Office, 2016). In specific terms, this increase means that approximately one third of the population will depend on a pension and a functioning social security system. If these systems fail, old-age poverty will become a real possibility. In addition to the issue of retirement provision, the labour market will also be affected by major changes in demographic structures that will intensify the competition for talent. Furthermore, due to the aging of the society, the skills required in a knowledge society will not be fully covered by existing skills, which will in turn make continuing education necessary to maintain the stock of human capital in the society. At the same time, the tax base will shrink for the state, as the income of retirees is generally not as high as that of the working population. This outcome will inevitably lead to a shortage of and more intense competition for government funds.

Knowledge society: The third development that should be mentioned here is the transition from a service society to a knowledge society. The creation of new knowledge is crucial to the success of an economy; in knowledge economies in particular (mainly Western countries), innovation accounts for 70%-80% of economic growth measured in terms of GDP (Information Society Commission, 2002). Thus, innovations are essential to the competitiveness of an economy, and universities play a key role in generating innovation (Stephan, 2012; Aghion, 2008). Governmental investment in research and development has therefore multiplied, and public spending on universities has skyrocketed, even in our own small economy of Switzerland: in the last 20 years, research and development (R&D) expenditure in higher education in Switzerland has more than doubled in real terms (OECD Statistics, 2018a). Increasing competition from Asia, and in particular from China and India, should not be ignored either. Over the past 20 years, China has increased its public and private research spending by a factor of 30. Due to this enormous growth rate, it is expected to overtake the US by the end of 2018 (it passed the EU back in 2015). These forecasts are based on the fact that China's average annual growth in research spending stood at 18% between 2000 and 2015, compared to 4% in the US. China is therefore preparing to become a leading scientific nation. China, with India in tow, may have joined the race late, but they are both going all out to catch up and overtake existing Western countries (OECD Statistics, 2018b; *Washington Post*, 2018). Thus, competition in research has intensified because of Asian countries, but new sources of research organizations, including platforms such as InnoCentive, are also playing an increasingly important role. NASA, for instance, posted a question on InnoCentive about solar flare prediction to which none of its engineers had been able to find an answer (Brynjolfsson & McAfee, 2016). After having been posted, the problem was quickly solved by a retired US engineer who had worked in a completely different field. The idea behind these kinds of platforms is quite simple: companies or organizations can make problems that they cannot solve by themselves internationally visible and thereby tap into an enormous additional source of human potential. An internet-enabled device, such as a cell phone, is all that is needed to use such platforms.

All the developments described above — globalization, demographic change and the knowledge society — intensify the competition worldwide and call for additional investment in tertiary education, since the generation of new knowledge is becoming increasingly important for growth and welfare, and because generated knowledge quickly loses its relevance. These challenges are putting European and North American science systems under pressure and sharply raising the costs of tertiary education, particularly in disciplines in which expensive research infrastructures dominate.

EUROPEAN AND NORTH AMERICAN SCIENCE SYSTEMS UNDER PRESSURE: WHO PAYS?

The situation faced by publicly funded research universities in continental Europe, with Switzerland as an example

Modern research universities can conduct highly competitive research only if they are given sufficient funds. In particular, expenditure is affected by the sharp and continual rise in the costs of modern infrastructure, especially in the fields of life sciences, natural sciences, medicine and high-performance computing. However, digitalization has also made a mark on other areas (e.g. the humanities) and has led to major costs resulting from the collection, management and storage of data previously not available in digital form. Until recently, universities specializing in arts and humanities were spared the expense of costly research infrastructures, but this is no longer the case. As a result, the vast majority of universities — at least in continental Europe — are finding it increasingly difficult to finance the additional expenditure through state contributions. This difficulty also applies to Switzerland, a rich country with a very stable funding system that is essentially based on three pillars. However, due to international competition and pressure, this system is now being pushed to its limits.

In Switzerland, due to its limited constitutional powers, the Confederation has little influence on higher education policy. The only area in which it has constitutional powers is the ETH domain that includes the two Swiss Federal Universities of Technology. The main bodies responsible for the 10 research universities are the so-called university cantons, which to a large degree finance their universities by themselves. The Confederation has a subsidiary allocation function in that it provides financial support to universities in the form of basic or investment contributions, or it makes funds available for special programs. The basic contributions are traditional financial subsidies and can be used freely by universities. In addition, as part of a horizontal financial equalization scheme (the Intercantonal Agreement), the universities receive funds from the non-university cantons for the education of the students from these cantons. This arrangement presents difficulties, however, as on the one hand, the university cantons are no longer willing to increase their contributions to the same extent as in recent decades, and on the other hand, the non-university cantons are no longer prepared to pay ever-greater contributions to the universities for their students. Thus, it should not be expected that funds will simply be increased to the extent desired by the research universities and their leadership.

The difficulties of sustainable research funding can be seen not only in Europe, but also in the US, where well-known research universities have amassed a mountain of debt to remain at the forefront of international competition. The University of California, Berkeley, a top-ranked, public research university, currently has debts amounting to $19.7 billion (University of California, 2018). The sky-rocketing costs and intense competition are a concern for all university presidents and can be described as a "race to the bottom".

A solution to the dilemma: "impact on society" or "third mission"

Recently, society's increasing investment in universities has led to a greater political focus on the topic of "impact on society". The debate, which started in the US and the UK, has also found its way into the politics of continental Europe. It calls for universities to implement their research results as quickly as possible to benefit society, create highly qualified work places, and, last but not least, generate additional income for the universities. This implementation requires universities to make a greater contribution to a region's prosperity — not only indirectly through increased educational returns, but also directly through research partnerships, patents, licenses and the formation of companies (spin-offs and start-ups) (Martin, 2012).

At the same time, the "third mission" is anathema to many university members who have been socialized in the publicly funded research universities of continental Europe and who fear for the independence of research and teaching at their own institutions. In addition, traditional research universities in continental Europe have never been accustomed to being held accountable for their impact on society. In Europe, the university governing bodies must make a large effort to implement the paradigm shift heralded by the "third mission" at their universities and to assuage people's fears.

Necessary investments to cope with the "third mission"

To cope with this political demand, university leadership is confronted with a number of new tasks. Essentially, these are awareness-raising among university members, training in additional skills, and providing appropriate resources. Without awareness-raising, young people are often unaware that they have the potential to start their own company to realize and commercialize their ideas. In addition, skills must be imparted for success-ful companies to emerge. These skills should be taught in various courses and workshops and by mentors. Last but not least, universities must make resources available, including legal advice on setting up a company, support for patent and licence management and help with the search for potential licensees or investors.

For the leadership of a university, the "third mission" means, above all, additional resources and investments in the first phase. Whether the univer-sities will actually be able to earn money from the spin-offs in later phases is not certain. Although some companies manage to make a major break-through and go public, this tends to be the exception. Most spin-offs become conventional small and medium-sized enterprises. Although the university does not earn money from them, the importance of these spin-offs for the university's region and for local politics should not be underestimated, par-ticularly as they may lead to the creation of high-quality jobs and tax reve-nues for the local governments.

However, the innovation pipeline can achieve a high degree of innova-tion only if the individual sections are correctly populated. For example, if not enough funds are invested in basic research in a country, not enough ideas will be produced. In addition, if new ideas cannot be translated into marketable products due to a lack of capacity in applied research or exper-imental development, the pipeline at the upper end will become blocked and result in too few innovations. The art of politics lies in making the right investments in the right places. Education economics has taught us that in technologically advanced countries such as Switzerland, government funds invested in research should first and foremost benefit basic research (Gersbach, Schneider & Schneller, 2008).

New Forms of Public Private Partnership for the "third mission"

In the past, we used to have a classic sequential innovation pipeline in which ideas from basic research were further developed in applied research before being tested in experimental applications; today, we see a change from the strictly sequential processes to parallel and ever-faster interactive processes (Gassmann, 2006; West & Gallagher, 2006). Specifically, this change means that basic research, applied research, experimental development, and appli-cation are linked via several feedback loops, greatly accelerating the imple-mentation of ideas. Thus, especially in medicine, translation research from the bench to bed becomes increasingly important and makes collaboration between different scientific disciplines and between basic scientists and cli-nicians indispensable for developing new therapeutic approaches. Currently, groundbreaking innovations in health care are not simply achieved in a research lab, but happen at the interface of academia, health care and industry. Based on this, our university has developed a public private partnership with Novartis, one of the world's leading pharmaceutical companies, and the university hospital. We founded the Institute of Molecular and Clinical Ophthalmology Basel (IOB), where basic researchers and clinicians work hand in hand to advance the understanding of vision and its diseases and to develop new therapies for vision loss (innovation). The setup of the institute is highly collaborative and interdisciplinary, and it is intended to increase the innovation rate based on the several feedback loops installed. Novartis is interested in this kind of research because innovation in ophthalmology has been slow for many years and because globally, the prevalence of eye diseases is constantly rising. Even today, there is no effective therapy available for most of them. In aging societies, disorders such as macular degeneration or glaucoma constitute a leading cause of disability and loss of independent lifestyle. Worldwide, and especially in Asia, myopia — or short-sightedness — is steeply increasing, with up to 90% of teenagers being affected in some regions. The IOB was set up as a collaborative organization to address precisely this challenge. It was established as an independent foundation, granting academic freedom to its scientists.

THE ENTREPRENEURIAL *MUST* OF THE UNIVERSITIES

A new form of leadership: dealing with politics and parliaments

The modern research university will face some major challenges in the coming years: international competition for top minds, international competition in research and development, and exponentially increasing research costs. No university can overcome these challenges alone: it depends on the support of the public and the politicians at the regional and national levels. This calls for new forms of collaboration and organization. As mentioned above, 70%-80% of the growth in prosperity in knowledge-based economies is attributable to new knowledge. Science policy is therefore becoming economic policy, and vice versa, for the first time in history. Both areas overlap and are interdependent. To a certain degree it automatically follows that university funding is no longer the central concern of only the educated middle classes, but that it is largely responsible for the development of prosperity in a country. However, if science policy is also becoming economic policy, universities must make efforts to create new alliances in politics, business and society. University leadership in the 21st century needs to become more political and entrepreneurial for the benefit of its institution and is required to obtain the necessary parliamentary majorities to develop further and conduct cutting-edge research.

Dealing with more stakeholders

At the same time, the university boards have a duty to broaden their funding base. In German-speaking countries, this cannot be done through tuition fees but only through private financing. Private money is generally acquired in two ways: fundraising and sponsorship on the one hand, and knowledge transfer and innovation on the other. Although they have already gained some experience with fundraising and sponsorship in the past decades, the field of "innovation" is still uncharted territory for many traditional universities in continental Europe. In particular, new forms of public private partnerships should be established to support the third mission of the universities and to increase the financial base for cutting-edge research. In addition, new forms of sharing infrastructure could be established among universities and corporations. In contrast to the 19th and 20th centuries, when traditional universities in Europe were integrated into the ministerial bureaucracies and when the university leadership only had to deal with the ministry, university leadership in the 21st century is challenged by new stakeholders and by the significance that the universities have for the welfare of society.

REFERENCES

Aghion, P. (2008). "Higher Education and Innovation", *Perspektiven der Wirtschaftspolitik*, 9, pp. 28-45.

Brynjolfsson, E. & McAfee, A. (2016). *The Second Machine Age*, Norton, New York, p. 83.

Eggenberger, J. (2015). "Fachkräftemangel in der Schweiz". Available online at: https://www.sko.ch/artikel/aktuelles/fachkraeftemangel [25 September 2015, version: 29 September 2017].

Gassmann, O. (2006). "Opening up the innovation process: towards an agenda", *R&D Management*, 36 (3), pp. 223-228.

Gersbach, H., Schneider, M.T. & Olivier Schneller, O. (2008). "On the Design of Basic Research Policy", *Economics Working Paper Series*, ETH, Zurich.

Haldemann, T. (1998). "Zur Konzeption wirkungsorientierter Planung und Budgetierung", in Budäus, D., Conrad, P. & Schreyögg, G. (Eds.) *New Public Management*, de Gruyter, Berlin, p. 195.

Information Society Commission (2002). "Building the Knowledge Society", Report to Government, ISC, Dublin.

Martin, B.R. (2012). "Are universities and university research under threat? Towards an evolutionary model of university specialization", *Cambridge Journal of Economics*, 36, pp. 543-565.

Müller, R.A. (1990). Geschichte der Universität. Von der mittelalterlichen Universitas zur deutschen Hochschule, Callwey, Munich.

OECD Statistics (2018a). "Higher Education Expenditure on R&D (HERD). Main Science and Technology Indicators Database". Available online at: stats.oecd. org [version: 26 September 2018].

OECD Statistics (2018b). "Main Science and Technology Indicators". Available online at: stats.oecd.org [version: 26 September 2018].

OECD Statistics (2019a). "Educational expenditures by source and destination". Available online at: stats.oecd.org [version: 20 September 2019].

OECD Statistics (2019b). "Gross domestic expenditure on R&D by sector of performance and source of funds". Available online at: stats.oecd.org [version: 20 September 2019].

OECD Statistics (2019c). "Social expenditure — Aggregated data". Available online at: stats.oecd.org [version: 20 September 2019].

Schenker-Wicki, A. & Olivares, M. (2010). "Innovation — Accountability — Performance: Bedrohen die Hochschulreformen die Innovationsprozesse an Hochschulen?", Die Hochschule, 19 (1), pp. 14-29.

Schüppel, J. (1996). Wissensmanagement: Organisatorisches Lernen im Spannungsfeld von Wissens- und Lernbarrieren, Deutscher Universitäts- Verlag, Wiesbaden.

Stephan, P. (2012). How Economics Shapes Science, Harvard University Press, Cambridge, MA.

Swiss Federal Statistical Office (2016). "Press release on population development scenarios in the Swiss cantons 2015-2045". [version: 12 May 2016].

Swiss Federal Statistical Office (2018). "Educational level of the population 1996-2017". [version: 29 March 2018].

Trow, M. (2007). "Reflections on the transition from elite to mass to universal access: Forms and phases of higher education in modern societies since WWII", in Forest, J.J.F. & Altbach, P.G. (Eds.) International Handbook of Higher Education, Springer, Dordrecht.

University of California, Office of the President (2018). "Debt summary". Available online at: https://www.ucop.edu/bondholder-information/debt-information/debt-summary.html [version: 26 September 2018].

Washington Post (2018). "China increasingly challenges American dominance of science", 3 June 2018.

Weiss, R. (2000). "Wettbewerbsfaktor Weiterbildung", Division, Dt. Inst.- Verlag, Cologne.

West, J. & Gallagher, S. (2006). "Challenges of open innovation: the paradox of firm investment in open-source software", R&D Management, 36 (3), pp. 319-331.

CHAPTER 8

Universities as drivers of societal development?

Michael O. Hengartner and Anna Däppen

R esearch and teaching have always been the two core missions of universities. But, central as they are, they only cover part of the spectrum of activities of modern universities. Indeed, urgent global challenges and the ongoing transformation of societies from agricultural to industrial to knowledge-based economies, have increased the public interest in profiting from academia also in other areas, including for example the transfer and exchange of knowledge (Ribeiro *et al.*, 2018). Universities are thus increasingly expected to actively promote interactions with industry and the society at large. These activities are often referred to as the "third mission" of universities (Etzkowiz & Leydesdorff, 2000).

The notion that universities can be agents of economic and societal development is, of course, not new; it had already emerged in Germany during the 19th century (Ribeiro *et al.*, 2018). History provides beautiful examples of the potential of universities to act as drivers of societal development, and many studies have confirmed the positive impact that can be generated by academic institutions (Blume, Brenner & Buenstorf, 2017).

THE THIRD MISSION

How broadly should this third mission be defined? That universities can contribute to the economic development of the surrounding community is undeniable. A recent study conducted by the League of European Research Universities (LERU, 2017) showed for example that the University of Zurich generated in 2016, directly and indirectly, more than €5 billion of economic activity and that almost 50,000 jobs depended, directly or indirectly, on the university. Furthermore, the University of Zurich holds over 300 active patent families and founds a spin-off company based on an UZH patent on average every other month, making UZH an important player within the regional innovation system. In recent years, observers worldwide have noted the significant influence of universities as knowledge providers on regional and national innovation and entrepreneurship (Blume, Brenner & Buenstorf, 2017). It is important to note that the fruitful transfer of knowledge and technology is not a one-way street, but rather a co-production process (van den Akker & Spaapen, 2017). Only then can innovations be successfully implemented outside academia. Hence, frameworks supporting an active exchange of ideas between science and society are of fundamental importance.

To reduce universities' impact within society to "simple economic metrics" (Benneworth, 2015) represents however a far too narrow view. While the promotion of economic development through cooperation with industry or the generation of spin-off companies is widely accepted and promoted, universities can also impact their communities in non-economic terms, including developments at the infrastructure and cultural levels. Thus, more and more, universities are expected to act as drivers of overall societal development by actively generating a variety of societal benefits (van den Akker & Spaapen, 2017). According to Paul Benneworth *et al.* (2019), there is actually a "myriad of ways in which universities contribute to changing the world by equipping civic society with new ideas, challenging injustice and reflecting on past failures, by creating platforms for silenced voices and supporting the development of better policies and better democracy".

As proposed by Chrys Gunasekara (2006), it might thus be helpful to differentiate between the different types of activities performed by universities. The previously mentioned knowledge capitalization of universities through activities such as licensing and spin-offs can be seen as a generative role that directly creates growth opportunities and which is mainly economic in nature. On the other hand, universities also play an indirect systemic capacity-building role, for instance by providing informed and unbiased analysis and information, thus contributing to the development of institutional and social capacities (Gunasekara, 2006). According to Gunasekara, this second role of universities can be characterized as developmental, going beyond the direct influence on economic growth.

It is not least based on the consideration that universities "can engage with and stimulate social innovation processes" (Benneworth & Cunha, 2015) that the University of Zurich (UZH) operates more than a dozen museums, botanical gardens and scientific collections, which are free and open to the public. They represent an important part of UZH's societal engagement, attracting more than 250,000 visitors per year.

UZH also offers a large collection of free lectures and panel discussions, including separate lecture series aimed at children, seniors and the general public. These activities generate an environment of openness where a broad variety of issues can be discussed and critically assessed. It is the right of free inquiry and freedom of speech, ultimately tied to the concept of academic freedom, which makes universities the predestined actors to foster openness and public engagement (Tierney & Lechuga, 2010). As part of its public lecture series, UZH regularly invites renowned personalities to present their views on a certain topic. Up until now, many important, but also controversial, thought leaders and politicians have spoken at UZH, among them Sir Winston Churchill, or more recently, the former president of the European Commission, Jean-Claude Juncker, Petro Poroshenko, (then) president of the Ukraine, or the Polish president Andreij Duda.

All these various activities of course require significant resources. However, we are convinced that they are a good investment, particularly since in Switzerland only about 20% of an age cohort go to university. By providing an open platform for discussion, UZH aims at contributing to the evolution of society as a whole by promoting a differentiated view on the world — something that is essential to the functioning of modern democratic and pluralistic societies.

Universities can also promote societal development through their core mission of teaching. By preparing their students to become informed and responsible members of society and by educating the thought leaders of tomorrow, universities are able to develop considerable transformative potential.

DEVELOPMENTS IN SOCIETY

Many important developments in society had their roots in student movements, one need only think of the far-reaching consequences of the protests in 1968. Universities can thus also facilitate societal development by encouraging and supporting student engagement. UZH has a long history of successfully promoting bottom-up student initiatives. In recent years, students at our institution have for example launched the Zurich sustainability week, an initiative to promote an ecologically friendly and sustainable lifestyle, or the Refugees@UZH Program, inviting refugees to attend lectures as guest auditors and eventually helping them prepare for a later application at UZH.

Last but not least, universities can of course influence society through the promotion of research on socially relevant themes. As free and independent institutions, universities have a unique capacity to analyse global challenges in all their dimensions and to offer solutions that take into consideration all relevant aspects of a problem. What is more, as places where many different perspectives meet, universities can provide a balanced view on potential risks and opportunities of developments such as technological change or digitalization. This consideration led UZH to launch a university-wide Digital Society Initiative (DSI) in 2016. DSI fosters interdisciplinary research on digitalization and promotes the dialogue with different stake-holders from inside and outside academia. Through their research, members of the DSI aim in particular at raising awareness of the effects and potential risks of a rapidly changing societal reality.

Of course, not every societal change is positive, and not every status quo is bad. Academic research can on occasion generate positive impact simply by acting as a stabilizing and integrating force within society. For example, the University of Zurich maintains a professorship of Romansh language and culture. Romansh, a descendent of Latin, is spoken by about 60,000 people living in a handful of valleys in the Swiss Alps. Although less than 1% of the Swiss population speaks Romansh today, it is one of the four official languages in Switzerland. Thus, although the small number of students speaks against it from an economic point of view, this professorship provides an important academic anchor for a language and a culture that represent an integral part of Swiss history and identity, the preservation of which is important for the cultural and national cohesion of the country.

TO SUPPORT OR TO DRIVE?

From the above, it is clear that universities definitively can, through their various activities, impact societal change. The final question that needs to be addressed is whether universities should act in a supportive role, helping society achieve changes that it deems worthwhile, or whether universities should aim to be in the driver's seat, set the developmental agenda for society and then spearhead these changes. While the latter would be intellectually attractive, it would, in our opinion, be counterproductive. The mission of public universities is to support society, not to boss it around, no matter how well-intentioned the bossing around might be.

This is not to say that universities never change society. But, ironically, history suggests that in many of the cases where universities did drive societal changes, these were not planned, but rather inadvertent side-effects of internal developments that were meant to only affect the university itself. As an illustration of this point, let us analyse two examples from the history of the University of Zurich (UZH), in which internal, "academic" decisions on how the university operates led to significant changes in Swiss society.

Being a country with few natural resources and an early industrialization, Switzerland became a comparatively early knowledge society and the establishment of institutions of higher education was seen as being of great public interest. The development of Swiss universities is in general closely linked to the development of the societies they are part of. This is particularly true for the University of Zurich, which opened its doors in 1833 as one of the first universities in Europe to be founded by a democratic state and not by a monarch or the church. In other words, UZH was founded "through the will of the people" and in response to public needs. The close relationship between the University and the community in which it is embedded explains why, at several points in history, university affairs gave inputs for lasting societal transformation. This was the case, for example, in 1839 when the appointment of the very liberal German theologian David Strauss to the Faculty of Theology of UZH caused great waves outside academia. The more conservative parts of the population who saw the old religious order endangered raised vehement protests against the appointment. On 6 September 1839, several thousand people stormed the city of Zurich, where a battle erupted between the protesters and the army, leading to 15 deaths and many injured. The liberal government, in disarray, was ousted and replaced by a conservative "provisional" government which held power for six years. The event was later referred to as the "Züriputsch", making the Swiss German word "putsch" an official German term to designate an uprising or coup d'état.

The graduation of female Russian student Nadezhda P. Suslova from the University of Zurich in 1867 is another example of how universities' actions can eventually initiate societal change. During most of the 19th century, women's rights to education were very limited throughout Europe. As a rule, only men were admitted to universities. There were a few exceptions, however. Following the lead from the University of Paris, the University of Zurich became the second university to allow women to study from the 1860s onwards. As there was no written law explicitly prohibiting the admission of female students, the president of UZH of the time took a pragmatic approach and allowed women to take up their studies at the University of Zurich. Over the following years, UZH attracted many young women, a large number coming from Russia, where previous reforms to girls' education had given women access to higher education, but without allowing them to pursue an academic degree.

Nadezhda Suslova was the first woman in history to formally enrol at UZH. In 1867, she graduated with a doctorate in medicine — the first woman ever to receive a doctoral degree in a German-speaking country. Suslova's pioneering achievement opened Swiss universities' doors to women. In 1872, merely five years after her graduation, women made up more than 30% of the registered student population at UZH, illustrating the lasting influence of Nadezhda Suslova's matriculation and graduation. Suslova's success initiated an irreversible — but originally unintended — development towards equal opportunities at Swiss universities and, through the professional, social and political activities of the female university graduates, also within Swiss society.

So how are we to answer the question addressed in the title of this contribution — are universities drivers of societal development? The answer is likely both a yes and a no. Universities' actions can indeed have profound influence on societal development. Some of them change society, others stabilize it or can even take it backwards. However, the two examples above also highlight the limited control that universities have on their actions' impact within society. To fully anticipate and control the consequences of university affairs and of scientific innovation is hardly possible. In most cases, only history will reveal the ultimate effects — be they positive or negative — of scholarly actions and decisions.

CONCLUSION

In summary, while the fundamental importance of academia's commitment to society cannot be denied, prioritizing societal impact at any cost and in every domain is likely not the most effective approach. In the face of limited financial resources and time, university leaders should set clear priorities, focusing on those areas where they can actively influence the outcome of their activities. Not surprisingly, these will often be areas corresponding most closely with the genuine strengths of academia, namely research and teaching. Therefore, we propose that universities should not strive to actively "drive" societal development. Rather, they should focus on their core business in the areas of research and teaching, thus providing the necessary basis for transformative scientific discoveries, education for qualified graduates and the means for successful science-society relationships. In short, it is by fostering excellence in research and teaching that universities can most effectively serve the interests of society and generate positive impact.

REFERENCES

Benneworth, P. (2015). "Tracing how arts and humanities research translates, circulates and consolidates in society. How have scholars been reacting to diverse impact and public value agendas?" *Art and Humanities in Higher Education*, 14 (1), pp. 45-60.

Benneworth, P. & Cunha, J. (2015). "Universities' contributions to social innovation: reflections in theory & practice". *European Journal of Innovation Management*, 18, pp. 508-527.

Benneworth, P., Fitjar, R. D., Fonseca, L., Manrique, S. & Nguyen, H. T. (2019). Special Issue Information. Online:https://www.mdpi.com/journal/socsci/special_issues/Universities_Contributions_Societal_Development (Accessed 23 July 2019).

Blume, L., Brenner, T. & Buenstorf, G. (2017). "Universities and sustainable regional development: introduction to the special issue". *Review of Regional Research*, 37, pp. 103-109.

Etzkowitz, H. & Leydesdorff, L. (2000). "The Dynamics of Innovation: From National Systems and 'Mode 2' to a Triple Helix of University-Industry-Government Relations". *Research Policy*, 29, pp.109-123.

Gunasekara, C. (2006). "Universities and associative regional governance: Australian evidence in non-core metropolitan regions". *Regional Studies* 40 (7), pp. 727-741.

LERU (2017). "The economic impact of the LERU universities". Online: https://www.leru.org/news/the-economic-contribution-of-leru-universities-2016

Ribeiro, B., Bengtsson, L., Benneworth, P., Bührer, S., Castro-Martínez, E., Hansen, M., Jarmai, K., Lindner, R., Olmos-Peñuela, J., Ott, C. & Shapira, P. (2018). "Introducing the dilemma of societal alignment for inclusive and responsible research and innovation", *Journal of Responsible Innovation*, 5: 3, pp. 316-331.

Tierney, W. G. & Lechuga, V. M. (2010). "The Social Significance of Academic Freedom". *Cultural Studies ⇔ Critical Methodologies* 10 (2), pp. 118-133.

van den Akker, W. & Spaapen, J. (2017). "Productive interactions: societal impact of academic research in the knowledge society", LERU position paper, March 2017.

CHAPTER 9

Pioneering Intellectuals and Innovation of Higher Education

Jaeho Yeom

And no one pours new wine into old wineskins. Otherwise, the new wine will burst the skins; the wine will run out and the wineskins will be ruined. (Luke 5:37)

CHALLENGES FOR HIGHER EDUCATION IN THE 21ST CENTURY

The 21st century is transforming the social modes of human civilization at an unprecedented scale and speed. Digitization is changing the world. The massive transformation of working behaviour, production systems, home automation, energy utilization, social and political systems, to name a few, are under way. This mirrors phenomena derived from the advent of electricity in the late 19th century. Now, due to exponential development in semiconductor technology, big data can be easily collected and accumulated in cloud computing systems, while information processing has reached lightning speed. Even though we still have to wait several decades for the full-scale revelation of the massive influence of AI on human civilization, the so-called "narrow AI", i.e. rule-based approach AI, is forcing humans to confront the deep learning revolution.

When AlphaGo defeated the Go world champion Lee Sedol in 2016, humans first noticed the shocking impact of AI on the future of human civilization. When narrow AI evolves into general AI, it will infiltrate all realms of human society. Man will be combined with machine in the near future. More than a decade ago, Ray Kurzweil (2005) predicted a new breed of human beings and of an unprecedented civilization in his book, *The Singularity is Near*, and is quoted as saying "I have set the date 2045 for the 'Singularity' which is when we will multiply our effective intelligence a billionfold by merging with the intelligence we have created." Yuval Noah Harari (2017) also claimed in *Homo Deus* that the difference between the present humans and the new humankind to appear around the 2050s is much greater than that between Neanderthals and present humans.

Applications of AI will be pervasive in the future society. Due to the prevalent use of AI in medical diagnosis and treatment, in addition to DNA analysis, for instance, the human life span will exceed well beyond 100 years. Human society will be more closely linked and networked. Cyberspace will be the venue for work, as well as a playground for humans. Traditional powerhouses will erode rapidly due to SNSs and the so-called cyber democracy. Powerholders will no longer be able to monopolize secured information, as we have experienced in the case of WikiLeaks. Everyone and anyone will be able to easily access information when needed. Time and space will be transcended in an unprecedented scale.

Education is not an exception, especially higher education, in this revolution of human civilization. The traditional way of teaching and learning shaped in the 20th century needs to be modified. Rapid economic growth and industrialization is greatly indebted to mass education, especially to higher education. Professional skills and specialized knowledge provided by higher education institutions have enabled college graduates to obtain better jobs, which has led to mass production and subsequently an affluent consumer society. Well-digested and highly specialized knowledge transferred from professors to students in classrooms has been applied effectively to the workplace. Transmitted knowledge and skills have allowed the maximum utilization of human capacity in the 20th century.

Starting from the 1970s, however, the introduction of computer systems and factory automation in the workplace began to change business operations as well as the production system on a profound scale. Human workers could not but yield to computerized and automated machines, so companies began to downsize human resources. In the 21st century, artificial intelligence has even rendered many human resources obsolete. The value of professional skills and knowledge owned by human workers is diminishing. Computer systems and artificial intelligence have belittled human capacity. Working hours have been shortened, and jobs need to be shared with other workers, as well as with machines. The gig economy is challenging the conventional, rigid labour system. Kai-Fu Lee predicted such a phenomenon in *AI Superpowers: China, Silicon Valley, and the New World Order* and stated that "...within 15 years, artificial intelligence will technically be able to replace around 40 to 50% of jobs in the United States." (Lee, 2018). Moreover, explicit knowledge in specialized fields will be rendered useless within a short period of time. Harvard researcher Samuel Arbesman empirically verified this claim in his book, *Half-Life Facts: Why Everything We Know Has an Expiration Date*. He found that half of the professional knowledge in a given field becomes obsolete within a decade. For example, the half-life period of physics is 13.07 years; economics 9.38 years; math 9.17 years; religion 8.76 years; psychology 7.15 years; and history 7.13 years (Arbesman, 2013).

The education system in South Korea (hereafter Korea) is standing at a critical crossroads. Facing a massive challenge to higher education, the Korean situation is even more serious because hard work, long study hours, rote memorization and cram schools used to be symbols of success in education. The unusual growth in the number of college enrolments has contributed to providing high quality human resources enabling the rapid economic growth of Korea. In the 1960s, only 6% of Koreans went to college. The number rose to around 12% in the early 1980s and has now reached 70%. Well-educated human resources equipped with specific knowledge and skills acquired from higher education have been and are still essential for the success of the Korean economy. From the least developed economy in the 1960s with a less than US$100 GDP per capita, the Korean economy has achieved prosperity as an economic superpower. As the world's 12th largest economy, with more than US$30,000 GDP per capita — a feat which has been achieved within five decades — Korea is one of the leading exporting countries of semiconductors, smartphones, home appliances, automobiles, steel, ships, refined oil and chemical goods, among others. However, the Korean economy is starting to face the limitations of rapid economic growth. Korean society now urgently needs to reformulate its system from a speedy catch-up economy to a front-running economy. Observers have claimed that one of the main challenges to this social and economic system reform of Korea is the transformation of its traditional educational system.

Due to this great challenge to higher education, we need to prepare for an innovative education system for the 21st century. The knowledge obtained from textbooks and classrooms is no longer effective for global economic competitiveness. Moreover, it cannot be monopolized by conventional higher education institutions as knowledge can simply be found, collected and accumulated in cyberspace. Memorized knowledge is not a power or capacity for professionals. Explicit knowledge is no longer a sufficient condition and is now limited to being a necessary one for human resources. It is the right time for us to change not only the content of specialized knowledge but also the methods of how to acquire it.

INNOVATION OF HIGHER EDUCATION FOR THE FUTURE

21st century higher education is confronting a massive transformation of pedagogy. Teaching methods have evolved in various unconventional ways in tandem with media technology development. The contents of knowledge, subject fields and majors are continuously changing and expanding. As a result, alternative educational institutions have drawn the attention of pioneering intellectuals and students. In the US alone, Singularity University, Minerva Schools and the Olin College of Engineering, to name a few, have aggressively innovated their curricula to nurture future leaders. They have experimented with pedagogy and knowledge content in unconventional ways. The Olin College of Engineering focuses its education on project-based learning of real-world problems. It emphasizes collaboration, interdisciplinary perspectives and state-of-the-art technology. Minerva Schools underscore diversity, multidimensionality and practical knowledge to face global uncertainties of the future. To benchmark the world's leading cities' problem-solving methods, students are required to spend one semester in different locations such as London, Seoul, Berlin, Buenos Aires, Taipei, and Hyderabad, in addition to San Francisco, where its main campus is located. Singularity University has also structured its curriculum so students can prepare to meet the world's most urgent problems. The commonalities found in these education systems are flexibility, a pioneering spirit and experimentation.

The conventional education system is facing a great challenge to deal with future global problems effectively. The traditional 4-year bachelor's degree may no longer appeal to students and the degree itself may not be worthwhile for a student's career. Students may prefer a nano degree or a microdegree for a specific subject which can be obtained in a short period of time to a bachelor's degree. A four-year residence requirement and high tuition and living costs are already a burden for college students. Rapidly changing new ideas, cutting-edge technology or newly emerging knowledge can no longer be obtained from conventional curricula. Instead, students may opt for short-term residence, intensive courses, flexible semesters, joint degrees, combined academic programs, internships and globally networked campuses.

The content of study in higher education should neither be limited to majors or traditional liberal art courses. Courses like design thinking, creative thinking, social problem solving, multidisciplinary courses and problem-based learning will replace content-based learning. In addition, team teaching, group discussion, experimental research and internships in the real world will substitute the simple instruction of concrete knowledge by professors in the classroom. In this light, professors will have to conduct classes in quite a different way from the traditional way of instruction. Flipped classes will prevail in higher education, in which students prepare for class by watching video clips of lectures outside the classroom in advance and later engaging in problem solving and discussion in the classroom. MOOCs, YouTube, TED and other online content are already being utilized. Attaining simple knowledge is no more important than mere thinking. Real world problems are more urgent than abstract ideas and concepts. Moreover, knowledge creation needs to be more valued than knowledge transmission in higher education institutions. In this sense, universities may compete with the business sector in creating state-of-the-art technology and knowledge. Companies in Korea, like Samsung Electronics, have more PhDs in its research labs than those at universities.

In 21st century higher education, we should not overlook the significance of the human relations capacity of students. A mature attitude and personality are important for future leaders. As more and more work is conducted through collaboration, human relations and group sensitivity have become more essential than the acquisition of specialized knowledge. In Korea, more than 30% of new employees in major companies are reported to leave their jobs within a year because they cannot endure the social conflict arising from human relations in the workplace, even though they were hired through intense competition.

Social responsibility and social value are other issues that higher education institutions should seriously consider. As the consumer reputation of corporations and of brands have an enormous impact on product sales, companies have begun to consider social responsibility in earnest. Business ethics and corporate responsibility have become non-negligible factors for the success and survival of corporations. Profit maximization can be a short-term goal for corporations, but if a company emphasizes the social value of the firm and its products, it can achieve long-term profit and growth. Consumers can easily surveil a corporate's activities scrupulously, because they can readily access relevant information in cyberspace. Additionally, collected information can be easily diffused among consumers through SNSs. If a company attains an unfavourable social reputation, it will detrimentally affect its sales. Not only business organizations, but governments, civil organizations and even universities are vulnerable to public criticism without exception. Thus, future social leaders will need to hone their social sensitivity and social responsibility.

VISION AND EXPERIENCE OF KOREA UNIVERSITY FOR EDUCATION INNOVATION

For education innovation, we need to modify our vision of student education which is based on the provision of professional knowledge. We need to expand our vision to encompass knowledge creation and social values. The key virtues for future leaders, whom we have to teach, are not limited to expertise and advanced knowledge, but should include embracement, external focus, clear thinking, imagination and courage. In short, future leaders educated in higher education institutions need to be social innovators with a broad vision. They should develop the capacity to solve social problems effectively. They can no longer secure life-time careers by simply carrying out given tasks. Rather, they have to identify problems incessantly and to discover unique ways to solve those problems.

Korea University's motto of education is liberty, justice and truth. Korea University has emphasized not only academic knowledge based on the value of truth, but also individual freedom based on the value of liberty and social engagement based on the value of justice. To apply such a vision to future education, it is necessary to broaden our vision from academic ability to other social values. Students need to acquire a pioneering spirit as liberal individuals in order to be future leaders. There is no right answer to many social problems so students should find solutions on their own. Challenging uncertainties, risk taking and seeking out unpaved roads are their missions for the future. In addition, they should develop social responsibility. In the future society, everything is interrelated. Man cannot live alone. In workplaces, modes of conduct are operated not individually but in a team. Individual excellence and survival of excessive competition will not guarantee success in the future.

With this vision in mind, Korea University has innovated its educational system in various ways. The following are some examples of initiatives it has recently implemented. From recruiting talented students to maintaining a knowledge creation eco-system, the challenges are pervasive. However, without such trials to advance education, the future of higher education in Korea and beyond will remain in dire trouble.

First, the admission system has been reformed. In Korea, high school graduates can enter college through several ways. The main way is based on the College Scholastic Ability Test (CSAT). Another is through special admissions such as an essay test administered by individual universities coupled with CSAT scores. The recent change in the admission process at Korea University focuses not on scores but on the attitude, problem-solving skills and discussion capacity of applicants. Not based on CSAT scores, but based on high school performance, applicants are selected in advance as candidates. Six admission officers evaluate six different areas of the applicant's records which varies from academic performance, to leadership, personality, extracurricular activities, social engagement and community service. They select three times more candidates than the admission quota. Professors and admission officers then intensively interview the selected applicants. They observe a one-hour group discussion among applicants and evaluate the quality of the applicant's discussion and problem-solving skills. In addition, four examiners ask questions based on a problem set given to the applicants and discuss issues for at least 15 minutes with each individual applicant.

Korea University has also recently changed the name of its Admissions Office to the Center for Talented-Student Discovery. It has exerted intensive efforts to select high quality students from all over the country. Evaluation based on academic scores was the conventional means for admissions. However, the evaluation system of in-depth interviews and discussion allowed applicants to enter based on other competencies and their potential to become successful leaders. This innovation was confronted with furious opposition from cram schools mostly located in Seoul, at which students take private lessons to obtain better CSAT scores. This innovation in admissions enabled public high schools to become more competitive. Almost 1,000 high schools can now apply for 3,000 spots, which consists of 85% of freshman enrolment at Korea University. This change has influenced high school pedagogy from rote memorization to discussion on various social issues in the classroom. High school teachers have begun to understand that creative thinking and ideation are more important than rote learning.

Second, Korea University has restructured its academic semester in a more flexible way. Traditionally, the academic year of Korean universities is comprised of two semesters starting from March and ending in February. A typical semester lasts for 16 weeks including mid-term and final exams. The Korean Ministry of Education defines one course as 48 class hours per semester. Most Korean universities run their semesters for 16 weeks, 3 hours per week and students usually take 6 courses in a semester. Korea University has allowed professors to organize their teaching flexibly within a limit of 48 class hours. For example, professors can organize their semester in 8 weeks at 6 hours per week, or in 10 weeks at 5 hours per week. This allows them to incorporate more discussion and problem-solving sessions. In this arrangement, it is also possible for world-renowned foreign professors to come and teach an intensive course.

Due to the flexible semester, professors can allocate time more effectively on research. They can also utilize the extra time to engage in globally networked research. For example, they can spend more than six months for research abroad every year. However, this comes with a requirement to reorganize the course curriculum. While teaching hours in the classroom can be reduced, time for discussion and problem solving such as through team projects needs to be increased. Professors also need to guide students in what they should prepare for the class by themselves in advance such as accessing video clips and required readings. The university provides support such as teaching fellows who lead tutorials and assist faculty.

In order to facilitate such courses, the university has set up an infrastructure called the NEMO (network module) lecture system using 5G broadband on campus. Lectures can be accessed by students on smartphones anywhere on campus so they do not need to attend the class in person. They can even download lectures on their PCs anytime within a week. Students, instead, must participate in discussion sessions and problem-solving sessions in groups led by teaching fellows or the professor. For instance, a NEMO course will consist of two 75-minute lectures and one three-hour discussion and problem-solving session a week for eight weeks in total for one semester.

Third, global leadership programs have been enhanced. Merit-based scholarships have been abolished and in their place, Korea University is now granting scholarships to global leaders. For instance, full scholarships are granted for students for an eight-week summer Chinese language program through which 100 students are provided with full tuition, living expenses and round-trip airfare to China. A similar program for Spanish is conducted in Mexico and for Japanese in Kyoto, Japan. Additionally, in 2015, Korea University established the Nordic-Benelux East Asian University Consortium. Professors and students can apply for a university-funded joint research project or for an academic experience project at participating universities.

In addition to more than 1,000 students going abroad as exchange students every year, Korea University has joined the Venice International University (VIU) global universities network. The VIU consortium is comprised of 18 member universities, each of which can send up to 20 students per semester to stay at VIU as exchange students. Curricula are determined one year ahead by an academic council organized by delegates from member universities. Each member university can also dispatch one visiting professor per semester. Most courses are related to global and current issues through which students can raise awareness in addition to academic study.

Fourth, Korea University has focused on research more rigorously. Most private universities in Korea heavily rely on tuition fees for their budget. Even though Korea University is a private university, the budget of Korea University from research funds far exceeds than that of tuition. As aforementioned, knowledge creation has become more important than knowledge transfer at higher education institutions in the 21st century. Now, Korea University is not competing with other rival universities, but is competing with Samsung, SK, Hyundai and LG, the top four business conglomerates in Korea, in order to produce creative knowledge and new technology. In addition, Korea University no longer relies on government R&D funds and is raising research funds from private enterprises for developing state-of-the-art technology. Korea University has made contracts through KU (Korea University) Crimson Enterprises to develop joint research prospects collaboratively. It selected 100 top enterprises which are leading the development of world-leading technology. University professors and company researchers form a joint R&D project, in which they closely communicate and consult to develop new technologies.

Korea University has also established joint programs with the business sector. One example is a master's program with SK Hynix. Students receive full scholarships and participating professors receive R&D funds from the company. Upon completion of the program, all graduates are recruited by the company. Another example is the Department of Cyber Security undergraduate program. Thirty students enter the department as freshmen every year and receive full scholarships from the Ministry of Defense and stipends from the university. These outstanding students won the championship at the DEF CON hacker convention in 2015 and 2018. Just like the military academy, when they graduate, they become public officers in cyber security agencies.

Fifth, Korea University has attempted to transform its campus into a knowledge amusement park. University campuses need to reformulate their spaces from knowledge transmitting classrooms to knowledge creation workshops. Just like the leading IT companies, spaces should become more flexible, comfortable and imaginative. At Korea University in 2015, the Pioneer Village (π-Ville) was constructed not as a building with classrooms but as an idea incubating workspace with a motto adopted from Albert Einstein, "Imagination is more important than knowledge." The four-storey building was constructed with used shipping containers. There, students organize diverse teams and rent space to develop ideas, suggest creative proposals for social problems and incubate venture businesses. Within two years of operation, more than 60 teams have successfully completed their missions, while several teams have actually started businesses. In addition, to meet students' demand for a space where products could be tested, the university recently opened several workstations. These Makers' Spaces are equipped with 3-D printers, worktables, cutting boards and resident technicians who help out with students' work.

Another innovative building on campus will be the SK Future Hall, which is a seven-storey building comprised of 28,000 square meters to open in fall, 2019. The main function of SK Future Hall is not teaching but knowledge creation. Thus, there are no classrooms in the building. It consists of only discussion rooms, carrels, living labs and a convention hall. Every floor has a small compartment for food and drinks just like a business lounge at the airport. This building embraces the future of education that Korea University envisions.

CONCLUSION

In the 21st century, universities will have to innovate higher education on a more fundamental level. The conventional way of education is no longer valid for the 21st century knowledge society facing the fourth industrial revolution. Everything including vision, function, pedagogy, classrooms, campus spaces and infrastructure, and academic system need to be reformulated.

What is best for students and society in the future should be the major driving force in innovating the system.

To prepare for the future society, students should be educated as pioneering intellectuals rather than as specialists of a certain field. Not only professional knowledge but the ability to incubate creative and innovative ideas along with social responsibility and a mature character are indispensable elements of a future leader's quality. Social innovation and problem solving will what they will have to nurture.

In order to educate future leaders properly, colleges and universities need to transform their academic system in more innovative ways. They should eradicate conventional academic bureaucracy. Path-dependent archaic inertia should be abolished in this paradigmatic change of human civilization. New pedagogies, new education systems, flexible adaptation, new visions and values for education, and innovative academic governance should be introduced for the future. Without such endeavours, higher education institutions will have to face more severe criticism from society. They may even be abandoned by students and alternative organizations or systems may emerge to solve such problems in their place. As new wine requires fresh wineskins, universities need to shed old ways and begin anew.

REFERENCES

Arbesman, S. (2013). *The half-life of facts: Why everything we know has an expiration date*. Penguin Group, New York, ch. 3.

Harari, Y. N. (2017). *Homo Deus: A brief history of tomorrow*. HarperCollins, New York.

Kurzweil, R. (2005). *The singularity is near: When humans transcend biology*. Viking Penguin, New York.

Lee, K-F. (2018). *AI superpowers: China, Silicon Valley, and the new world order*. Houghton Mifflin Harcourt, New York, p. 19.

CHAPTER 10

Towards Sustainable Development — The Role of Universities in Lifelong Education

Tan Eng Chye

In this era of globalization and rapid change, the notion that a university is a place solely to get a degree will, in time, be superseded by a realization that even after graduating from university, none of us can afford to stop learning. The learning profile and needs of society have fundamentally changed, and this will impact the education model and delivery of universities across the world.

TRADITIONAL UNIVERSITY EDUCATION

Universities are set up as academic institutions with powers to award degree qualifications. University education in many countries has broadly evolved from either the British model of early and deep specialization, or the American model which favours a broad-based approach to learning. Higher education institutions in both countries are recognized for their high quality and excellent learning environment; both countries have produced the world's best universities that consistently dominate the major university ranking tables.

There are however distinct differences between the two models: the British higher education model is characterized by a tutorial mode of learning, subject specification and a focus on independent study, whereas the American model offers a broad and general education covering a variety of subjects (often delivered through a core curriculum), with flexibility to choose (or change) majors in the senior years. Both education models have traditionally been degree-centric; universities are organized by departments, both to pursue disciplinary research excellence and to plan and deliver degree majors, minors or programs.

Universities award degrees. Concomitantly, stakeholders tend to evaluate the performance of universities in education using indicators that are degree-centric. Government agencies, for example, assess publicly funded universities on measures such as attrition rates, average time to degree, graduate employment rates, graduate starting salaries, student satisfaction with their degree course and learning environment, and so on.

But fundamentally, the role of universities in education is to develop people. A university education imparts knowledge, skills and attributes so that people can lead productive and meaningful lives, and contribute to society and the economy. Although the traditional model of university education is a degree-centric one, this will in time to come, shift to also encompass lifelong learning.

WHY LIFELONG LEARNING?

What is the impetus for lifelong learning in universities? There are a few irreversible forces at hand.

We are now living in the fourth industrial revolution, marked by emerging technology breakthroughs in a number of fields that include robotics, artificial intelligence, nanotechnology, quantum computing, biotechnology, 3D printing and fully autonomous vehicles. Klaus Schwab describes how this fourth industrial revolution is fundamentally different from the previous three, which were characterized mainly by advances in technology. The underlying basis for the fourth industrial revolution lies in advances in communication and connectivity never witnessed before. These advances and technological developments are disrupting almost every industry in every country, heralding the transformation of entire systems of production, management and governance. (World Economic Forum, 2016)

Evidently, there will be many sweeping changes to economies, industries and structures, affecting jobs and the nature of work. Technological, product and business cycles are observed to be shorter and sharper, and any change in this globalized and increasing interconnected world will permeate through borders and ripple out quickly. No place is spared from change.

For most individuals, the days of a single, stable career and retiring with a good pension are over. (Being academics, sometimes we may not be cognizant that the safety of a tenured academic career does not extend to other employment sectors.) According to the Bureau of Labor Statistics, (Marker,

2015) the average worker currently holds ten different jobs before age 40, and this number is projected to grow. Forrester Research (Marker, 2015) predicts that today's youngest workers will hold 12 to 15 jobs in their lifetime. Universities thus need to shift from preparing students for a career of a lifetime to a lifetime of careers.

There are also demographic trends at play. With an increase in global life expectancy, we will also be spending longer years in the workforce. In general, people are not saving enough, whether on their own or through voluntary retirement schemes (World Economic Forum, 2018). Coupled with a low-interest-rate environment and reduction in state pension provisions, it is difficult to achieve retirement adequacy early and people cannot afford to stop working. A *Washington Post* article (*Washington Post*, 2017) notes that in America, people are living longer, more expensive lives, often without much of a safety net. As a result, record numbers of Americans older than 65 are working — now nearly 1 in 5. In Singapore, in view of the aging population and low birth dates, the minimum retirement age has been raised from 55 to 62 years old. From July 2017, the re-employment age has also been raised from 65 to 67; employers must offer re-employment to eligible employees who turn 62, up to the age of 67. (Ministry of Manpower, 2018) This provides older workers with more opportunities to work longer and to support themselves.

We are also witnessing the rise of new work models such as self-employment, freelancing and remote work. For example, technological advances that directly connect buyers and service providers have enabled the gig economy to expand greatly in the past decade. The share of the US workforce in the gig economy rose from 10% in 2005 to nearly 16% in 2015. (NACo, 2017) These new work models are not employer-based; instead of relying on employers to provide continuous training and upgrading, individuals will now have to proactively take responsibility and ownership of their skills development.

The need to continually retool and reskill is already acutely felt by those in the workforce. At least 1 in 4 workers in OECD countries is reporting a skills mismatch with regards to the skills demanded by their current jobs. (World Economic Forum, 2017).

All these point towards the growing need for lifelong learning. It will no longer be possible to frontload and compress education into a four-year undergraduate degree program as, unfortunately, and, perhaps embarrassingly, we do not know much about tomorrow's jobs.

From a societal point of view, there is an impetus for universities to play a greater role in meeting lifelong learning needs. Beyond credentialing and facilitating labour mobility, *The Economist* (*The Economist*, 2017a) has warned that when education fails to keep pace with technology, the result is inequality. Without the skills to stay useful as innovations arrive, workers suffer, and if enough of them fall behind, society starts to fall apart. This is a scenario societies would want to avoid.

HOW ARE UNIVERSITIES RESPONDING?

The need for individuals to engage in lifelong learning throughout their careers is clear. But a World Economic Forum report has found that while the skills required for most jobs are evolving rapidly, adult education and training systems are however lagging behind (World Economic Forum, 2017). How universities, given their existing structures, can and will evolve to become effective providers of lifelong learning education, is not as straightforward.

The market is responding and innovating to enable workers to learn (often times while working) in new ways. Online offerings are making it easier for professionals to upskill or to learn new skills. Massive Open Online Courses (MOOCs) are typically self-pacing and allow employees to pursue academic interests in a way that fits their work and personal schedules. To assess and validate student progress, some MOOC providers administer periodic tests and charge for the award of credentials. Some universities now allow certain validated MOOCs to contribute credits to their degree program requirements.

But online learning is not without its challenges. Only a small percentage of enrollees complete their course. Notwithstanding, online learning has opened up a world of opportunities for both students and content providers. Some herald MOOCs as the greatest leap for education access. Given the large number of users, the absolute reach and impact of MOOCs are significant. One can now access courses offered by Harvard University or MIT, from anywhere around the world.

Some universities have started launching their full-fledged courses online. Georgia Tech's MOOC-inspired online master of science in computer science is a strikingly successful example. Tuition was set at US$6,630, about a sixth of the cost of an on-campus degree. The online course enrolment increased to 6,365 in Spring 2018, making it the largest master's degree program in computer science in the US and likely the world. (Inside Digital Learning, 2018) A single master's program from Georgia Tech substantially expanded the annual output of Computer Science masters graduates in the US.

Other educational market innovations include new ways of connecting education and employment. Udacity has launched a series of nanodegrees in technology-focused courses, designed in partnership with employers. General Assembly, a private, for-profit education organization founded in 2011, has campuses in countries throughout the world to teach entrepreneurs and business professionals practical technology skills. The company's curriculum is based on conversations with employers about the skills they are critically short of. It holds events where hiring organizations can see the coding work done by its students. General Assembly measures its success by how many of its graduates get a paid, permanent, full-time job in their desired field.

PREPARING STUDENTS FOR LIFELONG LEARNING

In this discussion on how universities can play a role in meeting society's lifelong learning needs, perhaps the most exigent and relevant task at hand is to equip and nurture existing students with the capacities, aptitude and attitudes that will allow them to engage in lifelong learning. This necessitates a critical stocktake of the undergraduate curriculum structure — it goes beyond obtaining a right balance of breadth and depth — and it entails curating a curriculum that hones future-ready skills and traits.

In curricular design, it may be instructive to note that a 2015 study by the Hoover Institution (Hoover, 2015) has found that people with a vocational education are more likely than those with a general education to withdraw from the labour force as they age. This pattern has been observed in countries that rely heavily on apprenticeship schemes like Denmark, Germany and Switzerland. This study has led some to conclude that people with specialized training may be less adaptable, and that a university education cannot solely be for the purpose of helping graduates to find work immediately, without consideration of helping people to adapt to change in the workplace. Universities must thus be careful about disciplinary or vocational over-specialization, and pay greater attention to helping students to adapt to a future of change.

Possibly for similar concerns, British institutions are often criticized for early over-specialization. The 1997 National Committee of Inquiry into Higher Education commissioned by the UK government recommended for all higher education institutions to work to achieve "a better balance between breadth and depth across programmes than currently exists". (Times Higher Education, 2010)

Breadth in the university experience is an important aspect that helps develop the capacity of an individual to learn, unlearn and relearn. Proponents of liberal arts education argue that broad perspectives are the best preparation for multiple career paths in a changing world. A liberal arts education gives students exposure to a broad range of fields. Students learn to work independently and in groups, how to write, express and communicate well, how to analyse, critique and defend arguments using a variety of tools, both quantitative and qualitative.

Strong foundations for lifelong learning cannot be underestimated. Companies like Google and Ernst & Young have cited that learnability is more important than other traits when recruiting employees (Business Insider, 2016). Eric Schmidt, Executive Chairman of Google, says the company seeks "learning animals", people who are naturally driven to learn on their own. These companies have figured out the key to keeping their teams at peak performance is to choose employees who are predisposed to learn and grow on their own.

As the nature and structure of work are changing, every university will have to review its undergraduate curriculum to chart a course that prepares students for a future of lifelong learning, according to the institution's priorities, structures and resources available. There is no tried and tested magic bullet model to adopt. Some UK institutions have launched new degrees that replicate liberal arts degrees offered in the US. University College London (UCL), for example, launched the Arts and Sciences (BASc) degree in 2012, where students create their own bespoke program incorporating both arts and sciences subjects, and study innovative core modules to enhance the link between disciplines, together with a foreign language and a job internship. (UCL, 2018) This degree programme is pitched at the best students, "those who see themselves as wanting a leadership position". (Times Higher Education, 2010) Other universities have taken the approach that one of the best ways to engage students is to encourage them to ask and explore the "big" questions and how ideas fit together and relate to life. The London School of Economics and Political Science (LSE) introduced a compulsory flagship and award-winning interdisciplinary course, called LSE100: The LSE Course, that aims to support the development of intellectual breadth. LSE100 uses important issues of public debate to motivate investigations of research methods and the need for academic thinking. Contrasting disciplinary approaches are examined in the small weekly classes, where students investigate the methodological choices underlying different approaches (LSE, 2013).

At NUS (National University of Singapore), the educational model that we had adopted several years ago was based on building "T"-shaped competencies. The vertical part of the "T" refers to a major or specialization that a student would need to learn in-depth knowledge; the horizontal part of the "T" refers to broad-based learning. To ensure that our students have strong foundations, we now need a thicker and broader horizontal base. Our students must now learn statistics, quantitative reasoning and computational thinking as knowledge in these areas is very critical for emerging areas of artificial intelligence and data analytics. This very "thick" layer of general education at NUS has been reinforced with many more of the new skills that we feel all our students would need to have when they go out into the working world, such as global orientation and adaptability, and industry experience. About two-thirds of our students go on internships.

Beyond writing and numerical skills, the horizontal component must also infuse future-ready skills that cannot be easily replaced by automation or robots. Social skills, which universities traditionally are not involved in, are an example. David Deming (Deming, 2017) has written about the growing importance of social skills in the labour market. Since 1980, growth in employment and pay has been fastest in professions that put a high premium on social skills. There is value in the ability to manage relationships well; people who can effectively negotiate the division of tasks between coworkers form more productive teams. If work in future will increasingly be done by contractors and freelancers, then the capacity for co-operation and negotiation will become even more important. At NUS, we have developed a "Roots & Wings" program which is now in version 2.0. "Roots" refers to personal skillsets like resilience and mindfulness and "Wings" refers to interpersonal skillsets. Through this program, students learn about empathetic communication and, hopefully, they become more effective when interacting with their peers and leaders in the future workplace.

On disciplinary expertise, NUS is now advising students that a "T" is not good enough; we are encouraging students to read a double "T", which in mathematical notation, is a Pi or π. With a π-shaped competency, one of the majors a student takes at NUS will be in Sciences, Technology, Engineering and Mathematics, or STEM, while the other major will be in the humanities or social sciences. This versatility will give our graduates versatility to skill up in either areas or in a multidisciplinary area, when needed, in the future.

Formal undergraduate programs span four years and, no matter how long you keep them in the university, students are not going to be able to learn everything they need to, because of the rapid rate of change. No university will be able to provide students with all the skills that are going to be needed 20 years down the road. Hence, beyond modifying the undergraduate curriculum to prepare students for a world of change and lifelong learning, universities must gear themselves up to meet the lifelong learning needs of their students, graduates and the community they serve. This represents a shift in thinking of the model of university education, which traditionally, is centered on pre-employment education and training.

CHALLENGES AND OPPORTUNITIES

Will universities be able to respond to the changing and evolving needs of society towards lifelong learning and continuing education?

Lifelong learning is not about accumulating degrees, but engaging in bite-size and timely learning to upgrade and learn specific skills. Yet, it has been said that the model of campuses, tenured faculty and so on does not work well for short courses. (*The Economist*, 2017b) Traditional university faculty have other priorities in long-term research work, and hence, academic institutions may struggle to deliver fast-moving content. Contrast this with non-academic institutions like Pluralsight, which uses a model similar to that of book publishing; it employs a network of 1,000 experts to produce and refresh its library of videos on IT and creative skills. These experts get royalties based on how often their content is viewed; its highest earner pulled in $2 million last year. Such rewards provide an incentive for authors to keep updating their content. (*The Economist,* 2017b) Universities are, however, not structured along such incentives. Tenured faculty are usually far more concerned (and rightfully so) with achieving breakthroughs in their research area and to build their reputation within the field, than with thinking about the vocational lifelong learning needs of their students.

On a more positive note, technology will bring about many new opportunities for universities to design and deliver lifelong learning programs. With flipped classrooms, constraints such as locality and scheduling no longer exist. Learning of materials can take place offline and physical class sessions can be allocated to discussion and problem-solving. Technology will also allow learning to become increasingly social and interactive. With MOOCs, the institution's potential for scale and reach to new learners is immense.

As lifelong learning course offerings will have to be developed to meet market and industry needs, a shift to engage in lifelong learning may bring academic institutions and industry closer, and, through the course of consultation and collaboration, the nexus between research, education and industry can be strengthened in a positive and mutually beneficial way. Industry developments can inform research, and vice-versa; education can be enhanced with industry relevance. Novel modes of industry training and internship may also evolve.

Some universities may opt to segregate lifelong learning and traditional undergraduate degree course offerings. NUS on the other hand, is experimenting with assimilating lifelong learners with undergraduate and postgraduate programs, in a mixed classroom setting. We believe that adult lifelong learners can enrich the classroom experience as they bring with them valuable life and career experiences and mature perspectives; lifelong learners bring an opportunity for diversity and cross-learning in the classroom.

So far, no traditional research-intensive university is engaged in lifelong learning in a concerted and comprehensive way, or as a core mission. NUS is perhaps bold and innovative in this regard, as our institution aims to be an important lifelong learning institution in Singapore and the region. In 2018, NUS initiated a Lifelong Learners' program, which is the first in any university around the world, where all NUS graduates will enjoy automatic enrolment into all of our continuing education programs for 20 years. By

2020, NUS is aiming to offer 20,000 continuing education places annually, and this can potentially benefit our nearly 300,000 alumni. The range of courses and modules will be comprehensive, but there will be an emphasis on offering skills-based industry-relevant programs.

In conclusion, lifelong learning presents tremendous opportunities for traditional research-intensive universities to contribute directly to a growing and pressing societal and economic need. While universities are evolving to become more engaged in lifelong learning, whether it be through dedicated continuing education units, or experimenting with Coursera and other MOOCs, or innovating their own models, we need to acknowledge that anticipating future trends, embodying the mindset of lifelong learning, and providing access to lifelong learning demands a complex system involving multiple stakeholders. This goes beyond universities extending the reach of their programs from being front-loaded on undergraduates to delivering educational options to students of all ages. A whole ecosystem comprising governments helping citizens to understand future job markets and the skills they will require, and financial incentives to support skills upgrading, employers that create work environments that support lifelong learning, are all necessary to bring about this societal shift to stay relevant and competitive through lifelong learning.

REFERENCES

Business Insider. (2016). "Simple quiz will reveal your current learnability level 3". [Online] Available at: https://www.businessinsider.com/this-one-trait-will-help-you-get-hired-by-google-2016-7/?IR=T/#simple-quiz-will-reveal-your-current-learnability-level-3. [Accessed 30 September 2018].

Deming, D. (2017). "*The Growing Importance of Social Skills in the Labor Market*", The National Bureau of Economic Research. NBER Working Paper No. 21473. [Online] Available at: https://www.nber.org/papers/w21473. [Accessed 08 January 2019].

Hoover. (2015). "General Education Vocational Education and Labor Market Outcomes over the life-cycle". [Online] Available at: https://www.hoover.org/sites/default/files/research/docs/15113_-_hanushek_schwerdt_woessmann_and_zhang_-_general_education_vocational_education_and_labor-market_outcomes_over_the_life-cycle.pdf. [Accessed 30 September 2018].

Inside Digital Learning. (2018). "*Inside Digital Learning*". [Online] Available at: https://www.insidehighered.com/digital-learning/article/2018/03/20/analysis-shows-georgia-techs-online-masters-computer-science. [Accessed 08 January 2019].

LSE. (2013). "*LSE100 recognised in Teaching Excellence award*". [Online] Available at: http://www.lse.ac.uk/website-archive/newsAndMedia/newsArchives/2013/02/LSE100TeachingAward.aspx. [Accessed 30 September 2018].

Marker, Scott. (2015). "How many jobs will the average person have in his or her lifetime?" *Linkedin*. [Online] Available at: https://www.linkedin.com/pulse/how-many-jobs-average-person-have-his-her-lifetime-scott-marker/. [Accessed 30 September 2018].

Ministry of Manpower Singapore. (2018). *"Responsible re-employment"*. [Online] Available   at: https://www.mom.gov.sg/employment-practices/re-employment. [Accessed 08 January 2019].

NACo. 2017. *"The Future of Work: The Rise of the Gig Economy"*, NACo. [Online] Available   at: https://www.naco.org/featured-resources/future-work-rise-gig-economy. [Accessed 08 January 2019].

The Economist. (2017a). *"Equipping people to stay ahead of technological change — Learning and earning"*, 14 January 2017. [Online] Available at: https://www.economist.com/leaders/2017/01/14/equipping-people-to-stay-ahead-of-techno-logical-change. [Accessed 08 January 2019].

The Economist. (2017b). *"Established education providers v new contenders — The return of the MOOC"*, 12 January 2017. [Online] Available at: https://www.economist.com/special-report/2017/01/12/established-education-providers-v-new-contenders. [Accessed 08 January 2019].

Times Higher Education. (2010). *"It's the breadth that matters"*. [Online] Available at: https://www.timeshighereducation.com/features/its-the-breadth-that-matters/414650.article. [Accessed 30 September 2018].

UCL. (2018). *"The Arts and Sciences (BASc) degree, Arts and Sciences (BASc), UCL — London's Global University"*. [Online] Available at: https://www.ucl.ac.uk/basc/prospective. [Accessed 08 January 2019].

Washington Post. (2017). *"Seniors' financial insecurity"*. [Online] Available at: https://www.washingtonpost.com/graphics/2017/national/seniors-financial-insecurity/?noredirect=on&utm_term=.d1c5f732b4af. [Accessed 3 June 2018].

World Economic Forum. (2016). *"The Fourth Industrial Revolution: what it means and how to respond"*, World Economic Forum. [Online] Available at: https://www.weforum.org/agenda/2016/01/the-fourth-industrial-revolution-what-it-means-and-how-to-respond/. [Accessed 08 January 2019].

World Economic Forum. (2017). *"Accelerating Workforce Reskilling for the Fourth Industrial Revolution"*, World Economic Forum. [Online] Available at: https://www.weforum.org/whitepapers/accelerating-workforce-reskilling-for-the-fourth-industrial-revolution. [Accessed 08 January 2019].

World Economic Forum. (2018). *"Challenges to preparing the world for retirement and 3 solutions"*. [Online] Available at: https://www.weforum.org/agenda/2018/01/3-challenges-to-preparing-the-world-for-retirement-and-3-solutions/. [Accessed 30 September 2018].

CHAPTER 11

Conversation is key — Universities and their responsibility for societal development

Sabine Kunst

INTRODUCTION

I n the winter of 1827/28, half of the city of Berlin — from the workers to the members of court society — listened to Alexander von Humboldt's legendary Cosmos Lectures, which form the nucleus of his great work. At each lecture, hundreds of listeners gained insight into the state of research at the time.

Last April, almost two centuries later, the President of Germany, Frank-Walter Steinmeier, alluded to the unique spirit of these lectures, officially opening a new series of Cosmos Lectures at Humboldt-Universität in light of the challenges facing research today:

"I believe that this spirit of the Cosmos Lectures is in fact needed far more today than it was in Humboldt's day. We are living in a time of great change. We are seeing ever faster and more powerful waves of technological disruption. We are experiencing tough global competition, which has long since ceased being just commercial rivalry, but has become political and systemic competition. Precisely at times like these — notwithstanding all the heated political debates on everything from migration policy to security policy — there is one thing we must not forget: the world's future, and our future prosperity, depend now more than ever on us working globally as equals, sharing scientific knowledge and viable solutions. If we in

Germany wish to continue to shape the future, rather than just to be driven along, then science and research must be a major priority in society" (Steinmeier, 2019).

Steinmeier's words make it very clear that academia and research must play a central role in society today. Research should not only ask questions of societal interest, but also enter into an active dialogue with society.

How is this dialogue achieved? And what is the role of a scientist or scholar at Humboldt-Universität today? Does it make a difference whether you pursue your research in the middle of a capital city or in a closed-off community of experts? Do we, like Alexander von Humboldt, want to make our knowledge and the results of our research public? And what are the framework conditions for the research production process and the criteria for success when it comes to transferring knowledge into society today?

Humboldt-Universität is a place where these questions should be asked with particular determination. And they should be answered with excitement and a willingness to experiment. When Wilhelm von Humboldt founded the Berlin University in 1810, his new, ground-breaking idea was to unite the two academic missions of research and teaching. He wanted to provide students with a well-rounded humanistic education at this "universitas litterarum". The concept spread quickly across the globe and resulted in a multitude of new universities being founded.

Two hundred years later, Humboldt-Universität is now expanding upon its founder's idea, pursuing a third mission that explores the reciprocal dialogue between research, teaching and civil society. There are many terms used to describe or define this mission and its many aspects: public engagement, open science, open access, knowledge exchange, knowledge transfer. What they all have in common is multidirectional communication and exchange, which are essential to the development of modern academia. Conversation is key.

But let me first ask where we stand in 2019 as a German "universitas litterarum"? It is our view that rational discourse as an essential characteristic of academia is endangered. It is in light of this that addressing socially relevant topics in research seems a necessary albeit not sufficient condition for universities to make a significant contribution to rational discourse today and thus attain sustainable societal development with a long-term perspective.

Launching its definition of Global Grand Challenges, the Organisation for Economic Co-Operation and Development (OECD) has made the point that issues such as (global) health, migration, financial (a) symmetry or green growth and sustainable development are, of course, reflected in today's research. And they have made it very clear that, in order to be profitable, this research should be conducted in exchange with non-academic experts in civil society.

Universities must therefore embark on a new dialogue with civil society, contributing and participating beyond mere academic publications and exhibitions. This requires appropriate formats that encourage a variety of stakeholders to enter into a dialogue with academia. But this path has to be marked by a thorough analysis of the implications and side effects it could have for academia and for the individual researcher. What incentives, for example, can we offer researchers with regard to their already challenging careers that would encourage them to embark on this path? What concessions are we willing to make to the comprehensibility and applicability of the research results?

I cannot give you a definitive answer to these questions. Instead, I would like to provide you with a concrete example of this dialogue with society. Humboldt-Universität fosters research both on societal issues and on public engagement. A new institution at Humboldt-Universität that exemplifies this is the Humboldt Lab, which is currently under construction and due to be completed by autumn 2019.

THE HUMBOLDT LAB

In the historical centre of Berlin, the Humboldt Forum will form a unique hub for art, culture, research and education with international appeal. In the near future, the rebuilt Berlin Palace, museums, the University and various event spaces will become a meeting place for people from all over the world — regardless of their background, age, education, interests, prior knowledge or preferences. In the Humboldt Forum, new forms of interaction are tested, a variety of cultural and social expressions can be experienced, and scientific and artistic ways of working are brought together. History comes alive in the present day.

The unique collections of the Ethnological Museum (Ethnologisches Museum) and the Asian Art Museum (Museum für Asiatische Kunst) will offer a comprehensive overview of the world's art and cultures spanning the ages as well as the continents. The Berlin Exhibition invites its guests to view its installations, multimedia projections and original objects, and thereby trace the developments and relationships, both past and present, which connect Berlin to the rest of the world. The Humboldt Academy provides educational services and overarching formats, as well as basic introductory programs accompanying and connecting the exhibitions and events. Moreover, it coordinates Humboldt Forum's research projects.

In the Humboldt Lab, Humboldt-Universität will provide research with a stage to render itself more accessible and comprehensible to a broad audience. In this 1000-square-metre Lab, ever-changing exhibitions and events will convey the role of research in everyday life. The Humboldt Forum is all about research approaches and cognitions, i.e. the methodological dimension of the emergence of new knowledge. Conversely, the University wants to show how academic work is conducted in various specialist disciplines — be it in the laboratory, while travelling or in archives — and what kind of questions researchers engage with around the globe. It is not only a showcase of academic achievements, but rather the visualization of cognitive processes in the history and present of academic practice, which also includes controversies, speculations, errors and limitations. The aim is to actively involve visitors, integrating them into a journey from the initial inspiration to the research breakthrough, and also to highlight the different approaches of the disciplines on this path of knowledge. For these creative design processes, experimental forms of exhibition presentation are being developed.

In this cooperation between academia and museums, the old idea of a "sanctuary for art and science" is brought back to the fore, which the Humboldt brothers sought to combine in the museum and the University in the 19th century. The city's cultural, research and educational institutions join forces to create a place of information, togetherness and pleasure.

Humboldt-Universität is participating in designing the Humboldt Forum in various ways. It is mediating research as a cultural practice of daring and precision, discussing socially relevant topics in academia, making disciplines and methods comprehensible as conscious, visual limitations, testing and developing new forms of academic communication, and demonstrating the importance of research and university for society as a whole. The Humboldt Lab will therefore exhibit core research, relying on the research focuses of Berlin's seven Clusters of Excellence (large-scale, collaborative research projects with specific research focuses, funded within the framework of the Excellence Strategy of the German the federal and state governments).

What is more, Humboldt-Universität will use the Lab as a venue to display its traditional collections. Among others, exhibits from the Sound Archives and the Computer Museum will illustrate the rapid technological transformation of our time — and even the visitors themselves will become part of a living artefact: a giant interactive display at the entrance of the Humboldt Lab showing a school of fish captures the movement of individual visitors and groups and displays them in an engineered yet artistic manner. This striking visualization serves as a transition into various academic discussions addressed in the exhibition. Making the visitors part of the exhibition symbolizes the participatory and dialogical approach of Berlin's Knowledge Exchange.

The exhibits will be moveable to guarantee individual views and multiple perspectives. Screens and displays will project different "layers" of information, e.g. on a large map of the world. The visitors will be invited to access them as individuals and to contribute personal opinions on the topics addressed. With this approach, the core topics of Humboldt Lab's opening exhibition, social challenges and climate change, will be just as dynamic, interactive and fast-changing as the underlying phenomena are in reality. In keeping with the tradition of the 250-year-old Cosmos Lectures, the Humboldt Lab will continue to promote and encourage vivid dialogue and exchange within innovative formats.

PERSPECTIVES

The Humboldt Forum is the logical continuation and expansion of our mandate as a university in the footsteps of Wilhelm and Alexander von Humboldt. So how did we arrive at this point? And what informs our process of developing a new third mission that fosters conversation between academia and society?

Universities are part of a living, developing society

All over the world, academic freedom is currently exposed to the scepticism of different groups. A lack of information can also trigger doubts about the credibility or even necessity of individual research projects. In addition to external attacks on academia, some threats are self-inflicted and inherent to the system. Instances like the replication crisis in bioscience research or the plagiarism crisis create doubt about the integrity and credibility of academic knowledge. Academia urgently needs to reclaim society's basic trust.

More often than not, criticizing academia also means criticizing the elite. A recent example of institutions reacting to the heavy criticism of social elites can be seen in the École nationale d'administration, a well-known French educational institution that is regarded as elitist and has been threatened with closure. It is not only in light of this that we, as research institutions, must concern ourselves with this part of the debate. We need to ask ourselves how universities are perceived by the general public and how we communicate. What kind of content can a university deliver as part of a knowledge exchange that actually interests civil society, and what does society expects from us?

As a result, educational institutions need to provide a sufficient flow of information that does not address the academic community exclusively. It must be ensured that civil society can formulate opinions based on open researched facts. This includes barrier-free access to and continuous quality control of the mediating instruments.

One of the institutions at Humboldt Universität that exemplifies this approach is the Berlin Institute for Empirical Integration and Migration Research (BIM), founded in 2014. BIM focuses on theory-based, empirical research that is always rooted in fundamental research, integrating a broad range of disciplinary perspectives — from empirical social sciences to religious studies, from linguistics to education studies. At the same time, BIM strives to achieve a systematic transfer of research into the public sphere, ranging from critical monitoring of political debates to events addressing the public and media interventions. In this respect, the BIM sees itself as an active observer of societal trends. It also serves as a bridge-builder between research and policy development, with many examples of BIM research directly shaping policymaking at the local and national level.

BIM is part of Humboldt-Universität, but it is also located in a city that is itself marked by exceptionally high levels of citizens with migrant backgrounds and that finds itself in close proximity to the policymakers of both the federal government and the state of Berlin. BIM brings all these realities into play and creates a unique new model of a scholarly institution that moves into the core of societal debates, simultaneously assuming the roles of analyst and fact-checker, moderator and coach, sounding board and consultant.

To provide an example, the Department of Integration, Sport and Football researches, consults and prompts public impetus regarding issues of achievement in sports and football in relation to social integration. On the one hand, the department distinguishes itself by conducting basic, theory-driven and empirical research regarding interdisciplinary research on integration, social capital and civil society. On the other hand, it conducts empirical, application-oriented research, which, for instance, includes evaluation studies on sports federations, sports associations and civic engagement. Research results are introduced into the academic community. At the same time, they serve as a source of empirical counselling support for actors in government, politics, civil society and the economy. Current research is thus linked to current debate, serving as the basis for a bidirectional knowledge exchange between university and society.

In short, we need to react to society's needs and developments by providing objective information and communicating it in a way that really reaches society. In order to achieve this, however, we must continuously assess and improve our communication formats and channels. I am thinking in particular of digital formats and technology-based knowledge transfer in order to interest younger sections of the population as well as new target groups. These target groups in return are estimated to supply researchers with new insights into contemporary requirements for teaching, learning and accessing knowledge in general.

In the field of teaching, "analogue teaching", with the immediacy of its knowledge transfer and the personal teacher-student relationship, remains highly attractive, perhaps precisely because it is analogue and therefore popular with students, increasing numbers of whom are "digital natives". Humboldt-Universität is therefore currently blazing the trail in the design of a very traditional and important authority within the University, the professorship. Together with the foundation Stiftung Humboldt Forum, we will fill a professorship for interdisciplinary curating with a decided focus on practical experience with an expert in exhibition practice.

All of these approaches aim to help us communicate better with society and open up to current issues.

Universities can't do it alone

The area of third mission has so far been rather generic in Germany, and researchers may often be unaware that their research contains elements that could be highly relevant to society. Often researchers involved with third mission projects working in one institution have no knowledge of each other and work in parallel. Accordingly, synergies and the potential for professionalization are not sufficiently exploited, there is only poor communication within the institution, and external communication through strategic bundling is effectively hindered.

However, we have to acknowledge that contemporary grand challenges, as well as their exploration, negotiation and communication, can only be successful in cooperation with other (educational) institutions. Universities are therefore advised to approach and design their third mission activities in cooperative formats. A collaboration that includes different types of institutions makes it possible to develop a particularly broad range of exchange formats with society. This is where strategy-building processes should be initiated to network the stakeholders more closely, sharpen the common goal of the third mission as a cooperative task and help ensure the effect of these joint efforts. By developing inter-institutional offerings and services, universities can become more attractive, accessible and visible. Of course, these offers will vary depending on whether it is a matter of cooperation between research and industry, between two educational institutions or between the university and the public sector. For all these scenarios, we must first develop our specific approach as a university and define the specific added value that we can create for our institution and for society here.

Cooperation may also help to identify one's own limitations and preconditions more easily in order to overcome institutional bias. German universities in general — and Humboldt-Universität in particular — have long been active in opening their doors through public events and a variety of participation formats. But it has not always been possible to create the awareness one might want to achieve and to develop new stakeholder groups. Building a third mission together with strong partners could mean generating added value in terms of the effect and effectiveness of this mission.

Humboldt-Universität is currently pursuing this path of cooperation primarily with the very successful museums that are an essential part of Berlin's incredibly rich cultural landscape. In the future, it will do so not only with the site of the Humboldt Forum mentioned above, but also with the Berlin National History Museum (Museum für Naturkunde Berlin), one of the world's most important research museums in the field of evolutionary, biodiversity and geoprocessing research. In order to make this cooperation tangible and accessible to the public, the museum and Humboldt-Universität are currently planning a joint science campus. The focus here is on two areas: citizen science and public engagement. Both are development areas for research institutions today, for a university even more so than for a museum.

In the close cooperation between these two institutions, we want to explore the extent to which the participation of civil society in research processes is a fruitful endeavour and how far it can be expanded. In the area of public engagement, we want to make concrete offers to researchers to develop new competencies in this field. Our aim is to strengthen the responsibility of research for and in society. In the long term, we want to increase the relevance and effectiveness of research, e.g. through more effective communication with and policy advice from researchers.

The Berlin University Alliance — the joint effort of Freie Universität Berlin, Humboldt-Universität zu Berlin, Technische Universität Berlin and Charité — Universitätsmedizin, currently a candidate in the German Excellence Strategy competition — is a further example of how to strengthen knowledge exchange by setting up a matrix structure and crossing institutional boundaries on a collaborative, city-wide level.

In summary, the platforms used for public engagement (and training on it) must focus on the joint development of experimental formats, academically, technologically and socially for the purpose of stronger mediation in analogue and digital formats — each focusing on immediate dialogue.

Universities are communities

In its most basic sense, the word "university" stands for the community of teachers and students. We would do well not to forget that. A good conversation needs both sides, society as well as university members — and this new focus on conversation and third mission is potentially challenging for a university. There is a balance to be struck when determining the relationship between freedom of research and teaching and new demands to open up academia to society.

Basic research, cutting-edge research that meets the highest standards as well as high quality teaching can only continue to exist productively if we do not overload it with an undifferentiated call for third mission or simply distract researchers from achieving optimum productivity. It must therefore be part of the claim "conversation is key" to make very precise and explicit distinctions, and not to expect each member of our institution in every phase of their academic work to engage in this conversation. A balance between academic and societal stakes and priorities is needed.

We should also focus more on our own research about a university's opportunities to engage in conversation with society. The discussion about "key performance indicators" and the question of criteria for the measurability of research impact is currently on the agenda on an international level. Humboldt-Universität is asking itself these questions with its newly founded interdisciplinary research institute, the Robert K. Merton Center for Science Studies. This interdisciplinary platform for research and teaching in science studies provides a place for exchange and cooperation for researchers interested in how science works and will include its object of research in the research process itself.

But if we want to reach as many researchers as possible with the idea that "conversation is key", we will have to provide attractive offers and incentives. Academic career paths are highly competitive, so we have to ensure that working with society also pays off within the academic community and among peers. Even if our researchers are intrinsically motivated to perform third mission functions, we still need to answer how we as institutions value, awaken and incentivize this commitment. And, above all, we need to determine how this translates into added value for the reputation of these researchers in a system in which reputation so far stems largely from within the academic community and where only the "academic impact" matters.

This calls for a change in political framework conditions, but also for a cultural change among the funding institutions for research and teaching. If they do not want third mission to be understood as one of many side-aspects to be ticked off when applying for funding, it must be provided with adequate resources. Last but not least, such incentives should include offers in the field of communication skills to help avoid conversation hindrances such as over-complex academic jargon. In our capacity as research management, we must therefore ask ourselves very precisely how we can help researchers and stimulate conversation.

In the future, this exchange will form part of academic education right from the very start. To this end, opportunities for continued education and training are necessary to make established academic staff aware of the significance of bidirectional transfer. At the same time, universities must familiarize themselves with appropriate tools and knowledge on how to engage with different target groups. Since aspects of the third mission, such as transfer and knowledge exchange, are currently considered potential performance dimensions in research funding and third-party funding, it is also advisable for universities to invest in their employees in order to prepare them early on for upcoming competitions. But, as a German foundation (the Körber

Foundation) just recently analysed for the Global University Leaders Council Hamburg, the third mission area has so far only been relatively weakly incorporated into the University's own formal administrative, managerial and organizational structures. At the same time, this study makes the important point that the political and legal framework conditions urgently need to be clarified so that the gap between the social demands placed on universities and the universities' ability to fulfil these demands set by its third mission does not grow any further.

I would like to stress that universities need to take into consideration more thoroughly the interests and requirements of their members in order to fulfil their function as places of rational discourse, production and development of socially relevant academic knowledge — even under changing conditions.

To sum up, the (third) mission of Humboldt-Universität will be to develop structures to best foster and improve dialogue and conversation with society. To achieve this goal, we will establish regional, national and international collaborations for the purpose of (joint) institutional policymaking. The University's aim is to establish Berlin as a capital of contemporary knowledge exchange in all its facets. Scholars from a wide range of fields are already researching contemporary societies and their individual challenges. Since the above-mentioned phenomena are not only relevant as abstract research questions for academic discourse, but should also, and more importantly, contribute to social development, they can't be considered and investigated without engaging in a direct dialogue with their subject: society.

REFERENCES

Steinmeier, F.-W. (2019). Opening of the Cosmos Lecture Series, part of the celebrations organized by the Humboldt-Universität zu Berlin to mark the 250th birthday of Alexander von Humboldt. Berlin, 6 April 2019. [Online] https://www.bundespraesident.de/SharedDocs/Downloads/DE/Reden/2019/04/190406-Kosmos-Lesung-Englisch.pdf;jsessionid=ED50AE7128D36B60DFB-F80F1B19E5786.2_cid371?__blob=publicationFile [Accessed 27 May 2019].

CHAPTER 12

Out of the Academic Echo Chamber: universities embracing innovation from unexpected places

Alice P. Gast

The sometimes-surprising trends of the last three years, from the rise of populism to the distrust of expertise, have led many to accuse the global elite, and academics in particular, of being out of touch with society and stuck in narrow "echo chambers". How can we ever hope to understand why millions feel alienated or "left behind" by elites and mistrust experts? If we are serious about maintaining the excellence of our universities and their societal relevance, the answer lies in a new mode of open engagement.

This is a critical time. We are shifting from a somewhat patrician era of university "community outreach" to a new age where we need to have true relationships and partnerships with local communities. We must change the way we interact with our neighbours and in so doing deepen our understanding and broaden our impact.

FOUNDING PRINCIPLES OF EUROPEAN AND AMERICAN UNIVERSITIES

The value of gathering teachers and scholars has been understood from the founding of the first universities in the 11th to 13th centuries. Institutions such as the Universities of Bologna, Oxford, Salamanca, Paris and Cambridge shared the goal of developing scientific and scholarly knowledge and transmitting it to others. This has been a constant whether the schools had religious or other cultural roots.

In the United States, the European model of the university was readily adopted. The first institutions — Harvard, William and Mary, and the University of Pennsylvania — were founded in the belief of the importance of education to society. The importance of universities to society continued to grow in the 19th century. The industrial revolution increased the need for an educated workforce and the institutional and societal impacts of the American Civil War created the need for new thinking about how these impacts should be addressed.

Several institutions with which I have been associated considered improving the world and educating citizens for the future as an important part of their missions. Lehigh University was founded in 1865 by Asa Packer to "Create an educated workforce to rebuild the nation..." contribute to the "intellectual and moral improvement..." Stanford University was founded in 1891 by Leland and Jane Stanford to create a coeducational, non-denominational, and avowedly practical university, producing "cultured and useful citizens".

Imperial College London's roots date back to the 1851 Great Exhibition. Prince Albert, Queen Victoria's Prince Consort, used some of the profit from that event to purchase 25 to 30 acres of land. Albert said: "I would buy this ground and place on it four Institutions corresponding to the four great sections of the Exhibition." He also said "I would devote these Institutions to the furtherance of the industrial pursuits of all Nations in these four divisions." He emphasized the importance of openness: "These Institutions must be open and common to all nations..."

OPENNESS AND THE IMPORTANCE OF THE MOVEMENT OF PEOPLE AND IDEAS ACROSS BORDERS

Openness and the ease of movement of people and ideas across borders have long been important to higher education. Exposing ourselves to new influences can fundamentally change what universities do. But it does not change what universities are for. We educate students to contribute to society. We pursue research at the forefront of discovery. We innovate for the greater good. That was true at Imperial College London and for our peers a century ago, and it is equally true today.

But we must find new ways to meet our traditional roles. Large parts of society feel like they are left behind and that universities are irrelevant to their lives. We need to find ways for even the most elite universities to be open to new people, new ideas and new partners. The universities that will thrive in the future will be those that are bold in their efforts to be build bridges to their local, regional and national communities.

Universities must be open to people, ideas and innovations from non-traditional places. We see this happening already. World-class universities are uncovering innovation from unexpected sources. They seek ideas from their communities, and work with them to develop those ideas. There is true two-way communications. This is not only a more effective form of "outreach", it is also a rich source of innovation. This matters to universities, to society and to innovation and enterprise.

RETHINKING ENTERPRISE

We have grown used to, and even expect, universities to be fountains of ideas and innovation. This is as it should be; the fruits of our research discoveries must be shared and developed to benefit society. New diagnostics, technologies, therapies and algorithms come from great universities every day.

The University of Cambridge, MIT, Stanford, ETH Zurich, Imperial and many others are natural hotbeds for world-changing startups by virtue of their clusters of great minds, research and education. Universities, and their leaders, can and must catalyse and direct growth in those innovation ecosystems.

For example, Cambridge Innovation Capital (FT, 2019), part-owned by the University of Cambridge, is investing hundreds of millions of pounds into that city's science and technology cluster. CIC enjoys "preferred investor" status for intellectual property coming out of the university, but more than half of its investments have no direct connection with the university. In oncology, CIC backs innovators such as the charity Cancer Research UK's lab, which has developed a "liquid biopsy" test for diagnosing lung cancer from blood samples and startups like Bicycle Therapeutics and Carrick Therapeutics.

A NEW GENERATION OF INNOVATORS

In recent years, we have also unleashed a wave of student entrepreneurs with their own ideas. Universities have supported them with mentoring, incubators, pitch competitions, seed funds and an "ecosystem". It is exciting to see student-driven ideas like Malav Sanghavi's ultra-low-cost baby incubator LifeCradle making a difference in the developing world or innovations like Elena Dieckmann's Aeropowder creating efficient insulation and packaging from waste chicken feathers.

At Imperial, this support and encouragement start on day one. As students register, they gain membership in the Enterprise Lab. Hundreds enter entrepreneurship competitions, like the Venture Catalyst Challenge or WE Innovate, a program especially for women entrepreneurs. Upon graduation, some stay within the university's ecosystem, for example through sponsorship for Graduate Entrepreneur or Startup Visas, or by growing their enterprises in Imperial College's White City Incubator.

We have rightly come to expect the unexpected. Students with the right combination of financial support and connections are opting to work for small fledgling startups instead of established corporates for their first job. These days one third of Stanford MBAs (Byrne, 2016) start their own company within three years of graduating, a quarter of MIT alumni (Matheson, 2015) have done so, with rapidly growing numbers of UK graduates taking this path. One challenge with this trend is that it is mostly available to graduates whose parents have the means to continue to support them. Others with great ideas need to take on less risky full-time employment. Schemes to support early-stage entrepreneurs is an important need for universities to help fill.

While many entrepreneurs, from London's incubators to Silicon Valley accelerators, are socially conscious and trying to make a difference in the world, their *weltanschauung* is limited by their life experiences. This is one reason why there are hundreds of startups and applications related to pizza delivery, music downloads and assorted others where they fulfil a societal need for people like themselves.

A rare few are breaking out of these "filter bubbles" and leveraging untapped potential. During the UK's economic downturn of 2008-13, Enterprise Ventures (now Mercia Technologies) achieved annual returns of 5-6% (FT, 2014) by deliberately investing in economically disadvantaged areas. It backed hundreds of small and medium-sized enterprises with hundreds of millions of pounds. These include Xeros, an AIM-listed, low-water-use washing machine manufacturer based in the relatively deprived South Yorkshire town of Rotherham. The polymer-bead technology, based on research from the University of Leeds, solves the kind of problem that too few venture capitalists consider, even when it's in their own backyards: how to deal with severe water shortages in places like England (Guardian, 2018) or California (Dimick, 2015). While it's not a unicorn, the economic and societal impact could be even greater, on its local community, and the world. It is certainly more compelling than yet another laundry app. (Pressler, 2014)

INNOVATION ON OUR DOORSTEP

What about those with bright ideas who are neither students, graduates nor researchers in elite universities? Our universities are often situated in communities where the residents have quite a different world view. People whose worries include paying the rent, avoiding violence and not falling ill are on a different scale to those of our academic stars. They need and deserve our support, and we can benefit greatly from their ideas. Their innovations are grounded in real, practical and important needs.

We see this in London's White City, an historically deprived community where Imperial College is developing a new campus. Imperial's Invention Rooms are providing maker spaces, hack spaces, students and staff to engage with our neighbours in pursuing their ideas. Recent inventions from the Maker Challenge for White City teenagers include fashionable new designs for hearing aids, a lightweight stabproof vest, a foldable skateboard and "sneaker speakers": trainers that can play music and convert kinetic energy into stored energy. The inventor of the sneaker-speaker thought that kids her age spend too much time on their headphones and they should be sharing their music and socializing.

These ideas come from, and can change, everyday lives. Ramona Williams, a local resident with visual impairment, struggled to use a baby buggy at the same time as a cane. She shared her ideas for a multi-functional baby buggy that would warn visually impaired users of hazards. Imperial biomedical engineering students and their tutor, Dr Ian Radcliffe, worked with Ms Williams to bring her ideas to life. The result, a "smart baby buggy", uses a combination of LIDAR — laser technology used in self-driving cars — and ultrasound sensors to warn users of oncoming hazards such as vehicles, pedestrians, curbs and drop-offs through vibrations in the handlebar.

At my former institution, Lehigh University, the Rising Tide Community Loan Fund provides micro and small business loans to entrepreneurs who find it difficult to obtain funds from traditional lenders. Otis McNeil, aged 18, (Rising Tide, 2019) had a vision for an energy conservation company that would work with local utility companies to help low and middle income families to weatherize their homes. But the banks would not support such a young entrepreneur, who, despite his strong contacts and experience, came from a poor neighbourhood. Rising Tide helped McNeil develop a business plan and gave him access to finance to get his startup off the ground. He now has eight employees, a 5,000-square-foot warehouse and a small fleet of trucks behind his profitable enterprise.

Lehigh economists are studying and enhancing the Rising Tide scheme to ensure that it will support all communities locally, including African American, Hispanic and Latino entrepreneurs who have traditionally been underserved in the region. It's another area where unexpected innovation, effective community relations and excellent research come together.

Breaking out of our comfort zone is hard. It requires investment, new partners, a long-term view and a continuous critical look at what we are doing. Universities are capable of doing that; some private sector players less so.

But that change is gradually, inevitably, necessarily coming as innovation's gatekeepers open up.

Renowned Silicon Valley tech accelerator Y Combinator is providing 1,000 low-income people across two US states with $1,000 a month (Weller, 2017) for five years. Colorado-based Common (FastCompany, 2018) is supporting social enterprises with a universal basic income (UBI) and mentoring. Crowdfunding platforms like Kickstarter and Crowdcube are bringing finance to areas that suffer from a venture capital drought. (UC Berkeley, 2017). In East London's Waltham Forest, Big Issue Invest (Big Issue, n.d.) is opening a multi-million pound hub to help marginalized young people enter the creative industries.

Last year, the UK National Advisory Board for Impact Investing, (NABimpact, n.d.) run by a distinguished group of City financiers, urged the government to invest £2 billion to seed billions more in private sector capital for economically deprived communities. Their focus on "impact investing" for sustainable, inclusive companies that provide a social good while generating healthy financial returns, rightly highlights an untapped locus of innovation. This is not a hypothetical. Massive institutional investors, like Bain Capital and Goldman Sachs, have been experimenting with impact investing for years.

CHANGING UNIVERSITIES, TRANSFORMING SOCIETY

We have the power to reshape what entrepreneurship means, not just in and around our campuses, but throughout society. One area where technological universities and our entrepreneurial partners have a common challenge is in the representation of women.

Fewer than one in 10 venture capital dollars go to companies with a female founder, despite studies indicating that female-founded firms make a higher return on investment (BCG, 2018). In this climate, visionary entrepreneurs are overlooked and markets remain untapped. As women are excluded from networks, the negative cycle perpetuates.

To fix this, women entrepreneurs need support at the earliest stage. This is why Imperial introduced WE Innovate (Imperial, 2018a): a contest and six-month program exclusively for female students, as they develop an early stage business idea, advance their leadership and learn entrepreneurial skills. WE Innovate has backed hundreds of female student entrepreneurs with mentoring, startup contests and exposure to investor networks. Some, like Pae Natwilai, (Imperial, 2017) a design engineer, founded Trik (Trik, n.d.) after developing intuitive tools and software for controlling drones. Pae's company, which is creating new jobs in the UK, has the potential to transform structural inspection in the energy and construction industries.

Investors are making a difference. Women-led businesses are the focus of Merian Ventures and Alexsis de Raadt St James has a track record showing that "unintentional familiarity bias in the venture capital industry creates opportunities for investors willing to back female founders".

The same spirit can be applied outside our own universities. We recently launched a pilot programme, "Agents for Change", which supports leadership, professional development and networking for women aged 18 to 64 in London's Hammersmith and Fulham, the borough that includes White City. We connect a diverse group of women with academic and business experts, as well as with each other, as they develop their communication, leadership and influencing skills.

Entrepreneurs are helping us to rethink what our faculty should teach, and how our academics can have a greater impact on society. When these innovators better represent society, their potential impact on universities and the world is amplified.

A RESPONSIBILITY TO ENGAGE: PUBLIC HEALTH

There are areas of public life where universities have a duty to engage at a deeper level, often because few other institutions share our convening power across sectors, and our ability to translate discoveries and innovations into real world uses. Public health is one such area.

Over the last generation, remarkable things have been achieved in global health, as WHO statistics show. Thirty years ago, 350,000 children a year contracted polio; in 2017, just 22 cases were reported (WHO, 2018a). Deaths from malaria have almost halved, from 839,000 in 2000 to 438,000 in 2017 (WHO, 2018b). Deaths from HIV-related causes have also fallen, down by 52% from the peak of the epidemic (WHO, 2018c).

These advances are accreting and accelerating. Global average life expectancy has increased by five years since 2000, the fastest increase since the 1960s. In the same period, we have seen deaths from malaria fall by more than 25 per cent, and HIV has become a disease that can be managed with treatment.

Meanwhile the health gap linking deprivation to poor health, results in the "inverse care law". In wealthy places, progress in healthcare has made once-terrifying illnesses, such as diphtheria, tetanus, whooping cough and polio, almost nonexistent through vaccinations, excellent clinical facilities and research hospitals. Yet tens of millions of children worldwide cannot access routine vaccinations or decent hospitals.

Social, geographic, political, economic and technological factors limit access to simple treatments. They are on the wrong side of an equation that medics have spent decades wrestling with: the so-called "inverse care law", (Lancet, 1971) whereby the availability of good medical care varies inversely with the local population's need for it.

This same inversion equation applies to almost all technologies: from smartphones to banking, electricity to food storage. But, with the right level of innovation and a new generation of entrepreneurs, this can change. Universities will become central to this process as an important convening power, bringing together public and private sector partners. National initiatives, such as Germany's High-Tech Strategy (BMBF, n.d.) and the UK's Industrial Strategy (GOV.UK, n.d.), can enhance this responsible stewardship of technology.

These quiet but profound revolutions have been driven by universities in collaboration with philanthropists, NGOs, governments and the private sector. Much of this progress has come from unexpected places. By necessity, we can't find answers in the lab. It needs more than traditional fieldwork, but a two-way or, more often, network of multiple conversations.

Universities can change the way we approach the significant challenges to maintain and improve the health of our diverse population. Never before have we had so much opportunity to do so. Technological innovations are breaking down barriers, bringing insights from abundant data and providing advances to areas such as global health, prevention and early intervention, food and nutrition, children's health and mental health. There is no better place to transform global and local community health than with world-class universities. They have strong collaborations across disciplines, and they produce research and graduates that improve health and wellbeing, through medical interventions and opportunities for prevention.

The new Mohn Centre for Children's Health and Wellbeing at Imperial's School of Public Health in White City shows one way we can do that. This state-of-the art hub for health and wellbeing research, education and community engagement is founded on the premise that all children deserve the best chances in life. By preventing chronic disease and infection in the early years of life, future generations have every opportunity to thrive and succeed. The Centre will support pioneering research, education, and community engagement that will improve the diagnosis, prevention and treatment of childhood illness on both a local and global scale.

Again, community participation can lead progress. A White City cohort study, following a group of children from birth into adulthood and old age, is among its first initiatives. By monitoring the health and lifestyle of participants over many years, Imperial will deepen understanding of childhood illness, and how disease in old age is connected to early-life experiences. This will also provide an unrivalled insight into the health of children and young people in White City, and allow for the development of interventions that address the health challenges they face. Such insights are crucial at a time when economically advanced countries like England face a growing gap in life expectancy (Imperial, 2018b) between the richest and poorest members of society.

FRUGAL INNOVATION, A HEALTHIER WORLD

The WHO estimates that about 80% of global blindness is avoidable. Imaging the retina can diagnose over 50 eye and whole body diseases — including glaucoma and diabetes — but current tools are slow, inaccurate, expensive and underused.

To eliminate blindness, the most important breakthroughs may not come from big science in academic or corporate labs, but from incremental improvements, and frugal innovations. When 90% of the world's blind live in the developing world, affordability of and access to treatments becomes the top priority.

When given the right support, students can spark unexpected and brilliant ideas that their more experienced professors may never have explored. Medical students Simon Rabinowicz and Uddhav Vaghela's invention and startup VUI Diagnostics could dramatically speed up the diagnosis of diseases that lead to sight loss.

Working in Imperial College's Advanced Hackspace, they rethought ophthalmoscopes, commonly used for examining the inside of a patient's eye, which are complicated to use, have a narrow field of view and are part of a cumbersome, slow process of eye analysis. They developed an affordable, simple and accurate plug-and-play retinal imaging tool, inspired by cheap consumer electronics and off-the-shelf features, like Bluetooth connections with smartphones and laptops, rather than specialist medical equipment.

Their tool is much faster than ophthalmoscopy, and can image ten times more of the eye, allowing for greater accuracy. Crucially, the device can be operated by those without clinical training, allowing retinal imaging to come to isolated regions without the need for advanced infrastructure.

REMEMBERING THE INDIVIDUAL

If you talk to the most visionary experts in global public health, they are imagining and working towards a world where health officials can design a malaria elimination strategy that works because it can be adapted to fit the needs of communities at the local level. Or where we can accurately predict — and protect against — the next pandemic. A world where levels of obesity and diabetes are falling, because policy-makers understand which interventions work, and which don't.

Traditional methods are insufficient to realize these ambitions. For example, working with, talking to, and being affected by, the people who live with outbreaks of infectious diseases like Ebola can make abstract problems concrete, and help academics to rethink their approaches.

One particular case that stays with Dr Nathalie MacDermott, a paediatrician, clinician and Ebola expert, is that of a 12-year-old boy in Liberia. "He was a double orphan. Not only had he lost his parents prior to the epidemic, but then his adoptive mother — who was a health care worker — contracted the disease and died. He himself contracted the disease but survived. I remember him clearly — he loved cookies and sausages, and was always asking for one of the two. Yet he developed a maturity beyond his years, and had seen things a 12-year-old never should. He was the only survivor from the treatment unit at that time, and so had watched everyone else around him die."

Although the medical team knew he had family members from his adoptive family, no one came to collect him. "He was left sitting on a step at the hospital, looking around for someone, but nobody came. There was simply no one to look after him. Not only had he lost his mother and survived Ebola, but he now was abandoned — and he was just one of thousands of cases."

A staff member took him in and helped track down the youngster's community, where he was left in the care of a community member, says Dr MacDermott. "I was desperate to know what happened to him, and spent most of the six months in Liberia trying to track him down, and find what happened to him — but I was never able to."

It changed the way Dr MacDermott thought about her research, which, she now believes, needs to be child-centred. She explains: "The mortality rate in children under one year olds from Ebola was close to 90%. Yet many of the medicines were not tested in this age group for a variety of reasons. We need to think carefully about how we include children in research in epidemics, as this would help us ensure they get the best possible care in the future — and the greatest chance of survival."

CONCLUSION

Universities were founded to help society, and their contributions to the world over past centuries are enormous. Our citizens are better educated and our workers are better trained. Our research discoveries and our innovative solutions to societal problems would make our founders proud.

The value of education and of universities used to be undisputed. That is no longer the case. Where once universities were considered an integral part of solutions to societal problems, their relevance is now being questioned. To some they are elitist institutions that do not understand the issues facing people who have not benefited from a university education.

Universities must adapt to these changing times. We need to rebuild trust and credibility. We need to demonstrate our relevance and importance to today's world. To do this requires that we expand our thinking and develop true partnerships with people beyond our staff, students and alumni. We have a tremendous opportunity to seek innovation beyond our rarefied campuses and echo chambers. We are already seeing signs that this is happening. I am optimistic that we will find great new innovations and great new partners in the world beyond our campus gates.

REFERENCES

BCG. (2018). "Why Women-Owned Startups Are a Better Bet" *Boston Consulting Group*, 6 June 2018. Online:
https://www.bcg.com/en-gb/publications/2018/why-women-owned-startups-are-better-bet.aspx

Big Issue. (n.d.). Online: https://bigissueinvest.com

BMBF. (n.d.). "The new High-Tech Strategy" *Federal Ministry of Education and Research*. Online: https://www.bmbf.de/en/the-new-high-tech-strategy-2322.html

Byrne, J. A. (2016). "MBAs Doing Startups At Nearly Four Times The Rate Previously Thought", *Poets & Quants* 26 June 2016. Online: https://poetsandquants.com/2016/06/26/mba-employment-reports-understate-startup-mania/

Business Insider. (2017). "One of the biggest VCs in Silicon Valley is launching an experiment that will give 3,000 people free money until 2022" *Business Insider*, 21 September 2017. Online: https://www.businessinsider.com/y-combinator-basic-income-test-2017-9?r=UK

Dimick, D. (2015). "5 Things You Should Know About California's Water Crisis", *National Geographic*, 6 April 2015. Online:
https://www.nationalgeographic.com/news/2015/04/150406-california-drought-snowpack-map-water-science/

FastCompany (2018). "Could this 'UBI for social enterprise' help fund business for good?" FastCompany, 23 May 2018. Online: https://www.fastcompany.com/40576169/could-this-ubi-for-social-enterprise-help-fund-business-for-good

FT. (2019). "Cambridge fund raises £150m in year's largest UK tech round", *Financial Times*, 31 March, 2019. Online: https://www.ft.com/content/27baa410-5245-11e9-b401-8d9ef1626294

FT. (2014). "Investing in deprived areas pays off for private equity fund" *Financial Times*, 18 November 2014. Online: https://www.ft.com/content/7fc2dea2-6c1c-11e4-b939-00144feabdc0

GOV.UK. (n.d.). "The UK's Industrial Strategy". Online: https://www.gov.uk/government/topical-events/the-uks-industrial-strategy

Guardian. (2018). "England at risk of water shortages due to overuse and leaks, report warns", *The Guardian*, 23 May 2018. Online: https://www.theguardian.com/environment/2018/may/23/england-at-risk-of-water-shortages-due-to-overuse-and-leaks-report-warns

Imperial. (2017). Online: https://www.imperial.ac.uk/news/183247/google-maps-structures-wins-innovate-uk/

Imperial. (2018a). "Imperial celebrates five years of pioneering WE Innovate women", Imperial College London, 25 October 2018. Online: https://www.imperial.ac.uk/news/188771/imperial-celebrates-five-years-pioneering-we/

Imperial. (2018b). "Poorest dying nearly ten years younger than the rich in 'deeply worrying' trend", Imperial College London, 25 November 2018. Online: https://www.imperial.ac.uk/news/189149/poorest-dying-nearly-years-younger-than/

Matheson, R. (2015). "New report outlines MIT's global entrepreneurial impact", MIT, 9 December 2015. Online: http://news.mit.edu/2015/report-entrepreneurial-impact-1209

NABimpact. (n.d.) Online: http://uknabimpactinvesting.org

Pressler, J. (2014). "Let's, Like, Demolish Laundry", *New York* magazine, 16 May 2014. Online: http://nymag.com/news/features/laundry-apps-2014-5/

Rising Tide. (2019). Online: http://therisingtide.org/custom-weatherization/

The Lancet. (1971). "The inverse care law", Hart, J. T., 27 February 1971. Online: https://www.sciencedirect.com/science/article/pii/S014067367192410X?_rdoc=1&_fmt=high&_origin=gateway&_docanchor=&md5=b8429449ccfc-9c30159a5f9aeaa92ffb; Trik. (n.d.). Online: https://gettrik.com

Weller, C. (2017). "One of the biggest VCs in Silicon Valley is launching an experiment that will give 3,000 people free money until 2022", Business Insider, 22 September 2017. Online: https://www.businessinsider.com.au/y-combinator-basic-income-test-2017-9?r=UK

UC Berkeley. (2017). "Crowdfunding expands innovation financing to underserved regions" University of California, Berkeley, 13 January 2017. Online: https://news.berkeley.edu/2017/01/13/crowdfunding-expands-innovation-financing-to-underserved-regions/

WHO. (2018a). Polio Q&A. Online: https://www.who.int/features/qa/07/en/

WHO. (2018b). World Malaria Report. Online: https://www.who.int/malaria/publications/world-malaria-report-2018/en/

WHO. (2018c). Global Health Observatory (GHO) data. Online: https://www.who.int/gho/hiv/epidemic_status/deaths_text/en/

WHO. (n.d.). "Blindness: Vision 2020 Factsheet", World Health Organization.

CHAPTER 13

The Meaning of Being Private in the Time of Great Change

Atsushi Seike

THE FOUNDATION OF LEARNING AND RESEARCH IS PRIVATE

Learning is, by nature, a personal activity. From an economic perspective, increasing our income by acquiring knowledge and skills through study is an "investment", and, furthermore, obtaining knowledge and learning new skills are forms of "consumption" that satisfy our curiosity. A necessary condition for this is to be able to study freely without being restricted by others. Therefore, state regulations governing academia are harmful, and as long as learning remains a personal activity, it is not something for which we can expect to receive public funding.

Carrying out research is also a personal activity. It is true that in totalitarian states, research is state-controlled and research activities may be considered "public affairs". However, in free states, research activities are "investments" in which we obtain research achievements that benefit ourselves, or "consumption" that brings about pleasure through the pursuit of truths based on genuine curiosity. Here too, the necessary condition is that we can freely carry out research without being restricted by others. Yet again, state regulations are harmful, and, once more, public funding should not be expected.

If this is the case, then activities of universities (schools), which provide a place for these individuals to study and carry out research must also be private. Universities provide a place where individuals can study freely and carry out research freely, as well as providing a service to support these activities. Hence, state regulations directed toward universities are harmful, and, in principle, public funding is unnecessary. In other words, studying and researching are personal activities, and universities, where these activities take place, are also private undertakings. Hence, it is fundamental that these establishments be private.

SOCIAL BENEFITS OF LEARNING AND RESEARCH

Yet, why is it that in many free nations, including Japan, national and public universities exist, and, furthermore, why is public funding also provided to private universities? This is because learning and carrying out research benefit not only the individuals doing these activities, but also benefit society as a whole. Learning and carrying out research not only generate investment return and consumption utility for individuals such as those involved in these activities as just mentioned, but through these activities, social benefits also arise.

Individuals developing their abilities through learning not only bring about an increase in their own income, but also improve the quality of available labour in a nation, thereby increasing the economic welfare of a nation as a whole. In addition, having intellectual citizens is also essential for the decision-making and management of a democratic nation, and in this sense too, it carries benefits for the entire country.

It goes without saying that academic progress through research brings about great benefits to all of society. The science and technologies that generated the material wealth of today are all made up of past academic achievements. Technological advancements resulting from developments in disciplines such as modern physics, chemistry and the life sciences, which began with the pursuit of truths brought on by the intellectual curiosity of individuals, have made it possible, for example, to manufacture useful yet inexpensive products, prolong lifespan and liberate people from heavy labour.

Because of the societal benefits of learning and research by individuals, citizens have a reason to give public funding to those who carry out such activities, and also have a reason to give public funding to universities where learning and research activities are carried out. For people who study at university, it goes without saying, but even those who do not go to university are implicitly agreeing to support these activities through taxes because, although it may be indirect, they will still benefit from the learning and research achievements of others.

Moreover, many of the social benefits of such learning and research will not be seen immediately, but will materialize over the long term. When it comes to research, those that respond to the societal needs of the moment are not the only ones that are useful to society, but rather, research whose usefulness is not yet known at the time enriches future societies. Research achievements driven by the individual curiosity of researchers develop basic research, and by researching the application of these, society becomes more prosperous.

Looking back on history, Newtonian mechanics, which became the foundation of modern science and whose principles are greatly enriching society today, had its origins in medieval European astronomy. This astronomy developed from the intellectual curiosity of astronomers who became interested in observational facts that could not be explained by the existing geocentric theory of the day. But, far from being appreciated by society at that time for their contributions, Galileo Galilei and others were tried by the Inquisition and socially persecuted. Only by guaranteeing that researchers who carry out such research and the universities to which they are affiliated have complete "academic freedom" can society reap long-term benefits.

This is also true for learning. Abilities that are acquired through learning are sometimes for the purpose of performing specific jobs (especially those at vocational schools, professional schools, etc.), but what is more important is the ability to think for yourself. Those who can think with their own mind and with logical thinking capabilities based on immense knowledge are the people who can perform their work while responding to the times when major technological changes take place, and, furthermore, are the people who can carry out the proper decision-making and management of democratic nations.

In other words, individuals must be able to conduct research activities freely and in accordance with their curiosity. Additionally, they must be able to think things through on their own without being biased and possess the ability to fulfil their work duties as well as their duties as citizens. This is the necessary condition for learning and research to produce social value. For this, at universities, which is where learning and research take place, academic freedom is indispensable, and this is the reason why citizens provide financial support in the form of taxes.

Problems arise when citizens do not necessarily fully understand this. People frequently are only interested in research whose outcomes will be immediately useful, and they also tend to insist on labour skills that will be of service right away. People have the tendency to feel that the only things worth spending their own money on are things that will provide immediate utility, and that it is unnecessary to give public funding to pursuits such as curiosity-driven research, the date of whose usefulness is unknown, or liberal arts education, for which there are no assurances that it will help accomplish our work at hand. In recent years in particular, this mindset seems to be getting stronger all around the world.

If this happens, the decisions on what kind of learning and what kind of research should be done will be in the hands of the people, who are the taxpayers, and will be narrowed in accordance with the thoughts of the government, which represents the people. Certainly, in today's democratic nations, no punishment will be imposed regardless of the kind of research or learning being undertaken, so in this sense, academic freedom has not been taken away. However, at universities and other institutions that rely on public funding for their survival and operation, it will become inappropriate to allocate resources toward research and learning for which receiving public funding is difficult. As a result, for those wishing to research or study these fields, academic freedom has in fact been constrained.

Academic freedom and the independence of universities, whose role is supposed to be to protect this, are at risk of being threatened not only in nations ruled under authoritarian dictatorships, but also in those with democratic governments. It is ironic that the democratic pressure from the very people who are supposed to reap social benefits through academic freedom has become the threat.

THE MEANING OF BEING PRIVATE

However, as mentioned above, individuals who learn freely and carry out curiosity-driven research are undoubtedly desirable for all of society. In particular, these individuals are not only indispensable for enriching the lives of people living today, but also for enriching the lives of future generations. In fact, on this point, I think that the meaning of the existence of private universities is extremely significant.

First let us make a comparison of national and public universities. Figure 1 shows the income structure of Japanese national, public (prefectural and municipal), and private universities. National and public universities rely on financial support from the national and local governments for about one-third of their income. In comparison, the proportion of income from tuition is half of this or less. Furthermore, if the ratio of public financial support to the total income excluding income from university hospitals (this is mostly offset by expenses for medical services) are calculated from the data in this figure, the percentage of public financial support is about 53% for national universities and about 59% for public universities, accounting for more than half of their total income. It is clear that they cannot get by without public financial support.

This reliance on others, such as the state, for their existence, carries the risk of threatening not only universities but also the independence of individuals in the first place. One of the first to point out this danger in Japan was Yukichi Fukuzawa, the founder of Keio University. In Fukuzawa's main work, *An Encouragement of Learning*, it is clearly stated that: "Those who lack the spirit of independence necessarily rely on others. Those who rely on others fear them. Standing in fear of others, they must fawn upon them. Their fear and subordination gradually become habitual; they come to wear faces of brass. They know no shame, and do not speak out on questions which call for discussion. In confrontation with others, they only know how to bow to the waist." (Fukuzawa, 2013).

It is exactly the same for the independence of universities. If you decisively rely on the state for your existence, you must ultimately comply with the wishes of the state. This is not just a danger in authoritarian states ruled by a dictator, but there is also a danger of governments chosen democratically by the people exploiting this. If citizens forget the social significance of academic freedom, there is always a risk of this occurring.

On this point, private universities have a great advantage. Their existence is preserved and managed through such sources of funds as the assets contributed by the founders who built the private schools with founding principles, tuition from students (their families) who approve of the educational policies of the schools, and donations from graduates who appreciated the education they received at the schools. Basically, they exist and are managed without public funding from the government, or without relying on it decisively.

Thus, the independence of universities is guaranteed by its financial independence, but, unfortunately, this condition does not exist in the case of national and public universities. Unless the people, that is, the taxpayers, at the very least give public financial support to the universities unconditionally, the independence of universities will not be guaranteed, but, in the case of private universities, the risk is comparatively less.

However, public financial support is also given to these private universities. This is because learning and research of individuals at private universities also have social value as mentioned above. In fact, as shown in Figure 1, even in private universities in Japan, public financial support accounts for 9% of the total income, and when income from sources such as hospitals are excluded, this becomes about 13%. In the case of Japan, the government also takes the independence and other such factors of private universities into consideration, and provides subsidies to private universities not directly but through the Promotion and Mutual Aid Corporation for Private Schools of Japan.

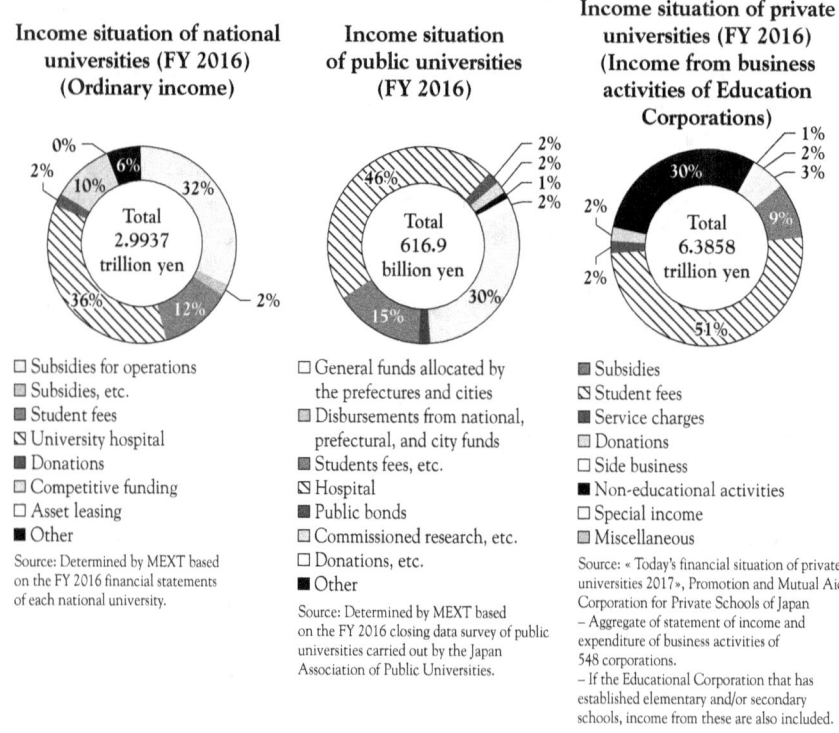

Figure 1– Financial situation of Japanese universities.
Source: "Summary of private university operating cost subsidies,"
The Promotion and Mutual Aid Corporation for Private Schools of Japan

Income situation of national universities (FY 2016) (Ordinary income)

- ☐ Subsidies for operations
- ☐ Subsidies, etc.
- ■ Student fees
- ◪ University hospital
- ■ Donations
- ☐ Competitive funding
- ☐ Asset leasing
- ■ Other

Source: Determined by MEXT based on the FY 2016 financial statements of each national university.

Income situation of public universities (FY 2016)

- ☐ General funds allocated by the prefectures and cities
- ☐ Disbursements from national, prefectural, and city funds
- ■ Students fees, etc.
- ◪ Hospital
- ■ Public bonds
- ☐ Commissioned research, etc.
- ☐ Donations, etc.
- ■ Other

Source: Determined by MEXT based on the FY 2016 closing data survey of public universities carried out by the Japan Association of Public Universities.

Income situation of private universities (FY 2016) (Income from business activities of Education Corporations)

- ■ Subsidies
- ◪ Student fees
- ■ Service charges
- ☐ Donations
- ☐ Side business
- ■ Non-educational activities
- ☐ Special income
- ▨ Miscellaneous

Source: « Today's financial situation of private universities 2017», Promotion and Mutual Aid Corporation for Private Schools of Japan
– Aggregate of statement of income and expenditure of business activities of 548 corporations.
– If the Educational Corporation that has established elementary and/or secondary schools, income from these are also included.

Private universities provide learning and research that benefit society in the same way as national and public universities. Whether learning and research are carried out at national or public universities or at private universities, there should be no difference in their value to society.

What is more, from the point of view that each private university has its own unique founding principles, it brings to society public benefits which national and public universities are hard put to do. In a time of great change like the present, it is important to increase diversity in various aspects of society in order to maintain the sustainability of society and, in this respect, private universities with differing educational and research policies based on different founding principles bring about diversity in the places where education and research take shape.

However, despite the magnitude of such social values, private universities should receive public funding in moderation. Private universities can protect academic freedom with greater strength, and, furthermore, can display diversity in education and research because they do not decisively rely on financial support from national and local governments. Although paradoxical, it can be said that because they are private, these universities can also protect their worth as institutions that create social benefits.

JAPANESE PRIVATE UNIVERSITIES

In today's Japan, private universities are the foundation of higher education. Table 1 shows the changes in the numbers of Japanese universities and university students from 1960 to the latest available for private universities and national and public (prefectural and municipal) universities. First, as can be seen from the most recent figures of 2018, there are currently 603 private universities where a total of 2,144,670 students are enrolled, accounting for 77% of all universities and 74% of all university students. Both the number of universities and number of university students greatly surpass those for national and public universities, accounting for almost three-quarters of the whole, and it is no exaggeration to say that in Japan today, the provision of higher education cannot be accomplished without private universities.

Table 1: Changes in the number of universities and university students.
Source: *Statistical Abstract (Education, Culture, Sports, Science and Technology) 2018*, Ministry of Education, Culture, Sports, Science and Technology

Year	Number of universities			Number of university students		
	Total	National and public	Private (%)	Total	National and public	Private (%)
1960	245	105	140 (57)	626,421	222,796	403,625 (64)
1670	382	108	274 (72)	1,406,521	359,698	1,046,823 (74)
1990	507	135	372 (73)	2,133,362	582,749	1,550,163 (73)
2018	782	179	603 (77)	2,909,159	764,489	2,144,670 (74)

Looking at the population of 18-year-olds, the general age when students start their studies at university, the number of students admitted at universities, and the percentage of students that were accepted at university from 1960 to the latest available: in 1960, when the economy began to grow rapidly, the number of students that were admitted at universities was 160,000, and of the population of 18-year-olds (2,000,000 persons), only about 4% advanced. However, in 1970, when the post-war baby boomers reached the age to attend university, the number of students that were admitted rose to 330,000, and the percentage of 18-year-olds (1,950,000 persons) that advanced increased to 17%. Furthermore, in 1990, when the children of the baby boomers advanced to university, the number admitted was 490,000, and the percentage of 18-year-olds (2,010,000 persons) that advanced was 24%. In 2018, the most recent year for which data is available, 630,000 students were admitted, and 53% of the 18-year-old population (1,180,000 persons) advanced. (*Statistical Abstract* ,2018).

Corresponding with this trend and, as can be seen again from Table 1, in 1960, there were 140 private universities with a combined student body of about 400,000 students. This grew to 274 universities with about 1,050,000 students in 1970, increased to 372 universities with about 1,550,000 students in 1990; the 2018 figures, the latest available, show that there are 603 universities with about 2,140,000 students. On the other hand, the number of national and public universities increased from 105 with about 220,000 students in 1960 to 179 with about 760,000 students in 2018, but, compared with the huge rise seen for private universities, it can be said that this growth is relatively restrained. The changes in numbers of private universities and students since 1960 are closely linked to the percentage of students advancing to university. The post-war expansion of higher education opportunities would not have been possible without the expansion of private universities.

The graduates of these private universities, growing in number, are active in various areas of the Japanese economy and society today. Table 2 lists the ranking of the top 10 universities at which CEOs of listed companies in Japan studied. Out of these top 10 universities, six are private, including first-ranked Keio University (which alone has 298 former students who are now CEOs), showing that an overwhelming majority of employees in business and industry are supplied by private universities. In the political world too, in the past quarter of a century, excluding 2009 to 2011, all prime ministers attended a private university, and, in addition, the percentage is also high among professionals such as medical doctors, lawyers and certified public accountants. Furthermore, there is an increase in the number of civil servants who attended private universities such as Keio University or Waseda University, a sector which was traditionally dominated by national university graduates.

Table 2: Ranking of the top 10 universities at which CEOs of listed companies in Japan studied. Note: There are a total of 3,708 listed companies in Japan. Source: Created using data published in *Yakuin Shikiho* (Executive Officers Handbook; Toyo Keizai Inc.), 2019 edition.

Rank	Name of university	National/public or private	Number of CEOs
1	Keio University	Private	298
2	Waseda University	Private	193
3	The University of Tokyo	National	192
4	Kyoto University	National	104
5	Nihon University	Private	80
6	Meiji University	Private	77
7	Chuo University	Private	71
8	Doshisha University	Private	59
9	Osaka University	National	56
10	Hitotsubashi University	National	51

On the other hand, research, traditionally, was certainly centered around national universities. Of the members that make up Research University 11 (RU11), a group of 11 major Japanese research universities, nine are national universities, and seven of these are former imperial universities. Keio University and Waseda University make up the two private universities in this group. In particular, research in the natural sciences, which requires extremely large research funding, is difficult to carry out at private universities where public funding is small, and, other than Keio University and Waseda University, it is not easy for private universities to carry out research at a level comparable to the nine national universities. However, in social science fields such as economics, the research capabilities of private universities are also improving and equals those of national universities.

In these ways, Japanese private universities play a notable role in providing higher education that produces the workforce necessary to sustain the economy and society. Furthermore, from a research perspective, national universities have the upper hand when it comes to research in the natural sciences, but in terms of both quality and quantity of research in the social sciences and humanities and social sciences, private universities also play an important role. This means that the Japanese people are also maintaining universities that provide social benefits through the relatively small invest-ments made using taxes. It also means that students and their families are bearing much of the expense.

BALANCING BEING PRIVATE AND RECEIVING PUBLIC AID

As described above, the existence of private universities brings great benefits to society, both as places where learning and research can be carried out freely, as well as the driving force that brings about diversity in education and research by providing education and conducting research based on their unique founding principles. Therefore, as I previously stated, there is suffi-cient rationality for the public to provide financial aid to private universities.

At the same time, as I mentioned before, private universities can main-tain academic freedom with greater strength and demonstrate diversity in education and research because they do not decisively rely on financial sup-port from the national and local governments. Therefore receiving more and more aid from the government is not necessarily better. Earlier, quoting the words of Fukuzawa, he said: "Those who lack the spirit of independence ... do not speak out on questions which call for discussion. In confrontation with others, they only know how to bow to the waist." (Fukuzawa, 2013). Certainly for those in academia, what is being sought is to say what needs to be said, at times even towards the government and the people, from an independent standpoint, and, in this sense, the independence of universities, guaranteed through financial independence, is extremely important.

The point is to maintain balance and move towards the establishment of a way in which to receive financial aid while ensuring the independence of private universities. Ideally, the aid should be provided in a way in which the discretion of the government is infinitesimally small. Implementing this through a democratic government is preferable.

When this is adapted to the situation in Japan, financial aid to private universities is provided through the Promotion and Mutual Aid Corporation for Private Schools of Japan in the form of subsidies to private universities as mentioned above, and, broadly, there are two types: general subsidies and special subsidies. The former, general subsidies, in essence are aid provided according to a standardized value based on factors such as the numbers of students and faculty members, and this value is multiplied by a subsidy rate set depending on the financial situation at the time to determine the amount of subsidy that will be provided. On the other hand, special subsidies are aid provided to private universities that offer programs or curricula or other such activities that are in line with the higher education policies of the government at the time, such as collaborations with industry, acceptance rate of adult students, promotion of international exchange and so on. The former is aid that is provided neutrally based on the content of education and research, while the latter accordingly guides the content of education and research of the universities that accepted the aid in a specific direction.

Of course, each university can make decisions on the programs, curricula, faculty recruitments and other such matters in accordance with their own policies on education and research. However, if they become more financially dependent on special subsidies, the result will be that they will eventually be guided by governmental policies. The government's intention from the beginning was to guide each university in accordance with its own policies.

Furthermore, another problem with the government's special subsidies is that not only are they just financial aid, but they are aid that also affects the branding of the university. For example, if a university is not receiving a special subsidy to promote internationalization programs, this university's efforts on internationalization may be seen as being inferior. However, each university is promoting internationalization in their own way that conforms to their founding principles, but those trying to internationalize in a way different to that recommended by the government are not in a position to receive the special subsidy. If, in order to develop high-quality international exchange, one university decides to establish a highly integrated program for home and international students by limiting the number of exchange students at a time when the government's policy aims to increase just the number of international students, this university will not be eligible to receive the special subsidy.

Because the financial aid is provided by the government, its use must certainly be in line with government policies. However, as I have repeatedly emphasized in this paper, I think that the use of the aid should not be confined to a limited range nor to short-term goals. The independence of private universities brings about diversity in education and research, and contributes to research achievements and the nurturing of personnel that are necessary for society in the long-term. Thus, guiding private universities in a uniform direction through special subsidies undermines the societal meaning of the existence of private universities.

At least from the viewpoint of securing its independence, which is at the source of the societal meaning of the existence of private universities, financial aid from the government should be in the form of general subsidies as much as possible. As shown in Table 3, the breakdown of financial aid provided to private universities as of 2018 is 269.7 billion yen for general subsidies and 45.6 billion yen for special subsidies. These respectively account for 85.5% and 14.5% of the total, and there is an increasing trend in the percentage of special subsidies. Returning to the original principles of private university aid, all governmental financial aid should in principle be general subsidies. Special subsidies should be external additions, and not be included as a part of the entire financial aid package.

Table 3: Changes in aid to private schools.
Source: "Summary of private university operating cost subsidies,"
The Promotion and Mutual Aid Corporation for Private Schools of Japan.

Unit: billions of yen

Year	Total	General subsidies (%)		Special subsidies (%)	
1975	100.7	99	(98.3)	1.7	(1.7)
1989	248.6	225.9	(90.9)	2.3	(0.9)
2011	320.9	281.2	(87.6)	39.8	(12.4)
2012	318.7	279.3	(87.6)	39.4	(12.4)
2013	317.5	278.3	(87.7)	39.3	(12.4)
2014	318.3	276.2	(86.8)	42.2	(13.3)
2015	315.2	271.1	(86)	42.2	(14)
2016	315.2	270.1	(85.7)	45.1	(14.3)
2017	315.2	268.9	(85.3)	46.4	(14.7)
2018	315.3	269.7	(85.5)	45.6	(14.5)

THE MEANING OF HAVING FOUNDING PRINCIPLES

In this paper, I have discussed the societal meaning of the existence of private universities. This is decisively dependent on each private university having its own educational and research policies, that is to say, the universities being managed independently. And, in the case of private universities, their distinctive policies on education and research are nothing more than the way in which their founding principles are implemented such that it applies to present day circumstances.

For example, in the case of Keio University, the founder, Yukichi Fukuzawa, established the university with the goal of nurturing "independent individuals" who are capable of living their lives without relying on the government or other public bodies, which is also referenced in the above quote. Over 150 years ago, in a time when Japan's independence was threatened by Western powers, Fukuzawa thought that first and foremost, each and every individual had to possess a spirit of independence to protect the independence of the nation, and took the decision to put this ideal into practice. Under the hereditary class system that existed up to that time, the Japanese people just followed in their parents' footsteps to make a living and lived in accordance with the rules set forth by the government. In a time when this manner of living was still the norm, he advocated the need for the people to build a life for themselves on their own, not leave everything up to the government, and protect the independence of the nation with a sense of ownership.

In *An Encouragement of Learning*, Fukuzawa writes that "... when the people of a nation do not have the spirit of individual independence, the corresponding right of national independence cannot be realized." (Fukuzawa, 2013). Therefore, he came to the conclusion that for individuals to possess the capability to independently make a living and have the ability to make judgements as members of the public, studying and learning are indispensable, leading him to establish Keio University.

Furthermore, he decided that this learning should not have its roots in the teachings of Confucianism that was mainstream in Japan up to that point, where people memorized things said by celebrated people of high social status as if they were golden rules, but rather in "science" that requires logical and empirical thinking. And, with this science as a foundation, he declared that of all the skills independent individuals need to possess, the development of "public knowledge", as defined in "... the ability to evaluate men and events, to give weightier and greater things priority, and to judge their proper times and places; let me call this public knowledge" (Fukuzawa, 2008). Fukuzawa's founding principles, including this insistence on independence, logical and empirical thinking, and public knowledge, have not faded and are still of importance today and are highly honoured at Keio University.

Although the current situation of Japan may be different from what it was back in Fukuzawa's time, the circumstances surrounding the country in terms of internationalization are becoming more and more uncompromising. Under these conditions, the competence of independent individuals to perform work as well as their decision-making abilities as members of the public are becoming even more significant.

In a time when we are faced with issues such as rapid population ageing, technological innovations and globalization, it goes without saying that our well-being and our potential as citizens can only be attained by improving our abilities to think scientifically and polishing our public knowledge such that we can determine what is important at any given moment in time. At the end of the day, for private universities, founding principles are the essentials, the alpha and omega. Even for the problem of education and research being guided by special subsidies, which I discussed earlier, if each private university has the option to adopt or forgo special subsidies based on their own founding principles, then it will also mean that private universities will not lose their diversity. The brand of a private school is not determined by the government, but through the endorsement of the people who empathize with the founding principles of the school and find attractive the distinctive educational and research activities of the school that is based on its own founding principles.

The uniqueness of private universities will be maintained if each and every private university constantly revisits their founding principles and reflects on its meanings. What financial independence of private universities means is having assurances that these universities can be operated under their own management to implement their individual founding principles in a way that makes sense in today's world.

REFERENCES

Fukuzawa, Y. (2008). *An Outline of a Theory of Civilization*, translated by David A. Dilworth & G. Cameron Hurst III, Columbia University Press, New York.

Fukuzawa, Y. (2013). *An Encouragement of Learning*, translated by David A. Dilworth, Columbia University Press, New York.

Statistical Abstract. (2018). Statistics, (Japan's) Ministry of Education, Culture, Sports, Science and Technology, Tokyo.

CHAPTER 14

The Challenges of a Liberal University

Pratap Bhanu Mehta

The modern Indian University dates back to the establishment of the three universities in the Presidency towns of Calcutta, Bombay and Madras in the middle of the 19th century. Indian higher education has taken diverse forms since then, from the prestigious IIT's to the 500-odd public universities. (Mehta & Kapur, 2017).

This short paper reflects on the three central challenges in building a Liberal University in the context of Indian higher education. The debate over the nature and character of a liberal university acquired its full vigour in India at the turn of the 20th century. One of its most succinct expressions was the Convocation address given at the University of Mysore in 1918 by Sir Ashutosh Mukhherjee, Vice-Chancellor, Calcutta University. The University of Mysore was the first "liberal arts" university set up in a princely state in India. Sir Ashutosh Mukherjee was a pivotal figure in the transformation of Indian Higher Education. He was instrumental in bringing the Humboldtian idea of a research university to India. This convocation address was remarkable in the way in which it prefigured many of the challenges of setting up a liberal university in India.

Mukherjee begins his address by raising the question of what is a University? He writes: "They have from time to time asked what a University is and found themselves at sea. Is it a set of fine buildings? Is it an education institution which has beneficent patrons and has secured the gift of a million? Is it an aggregate of the Four Faculties? Is it a scholastic guild? Is it a society of masters? Is it an assembly of students? Is it an examining body authorized to grant degrees? Is it a corporation of individuals who investigate the unknown, but neither teach nor test? Is it an association of teaching institutions without a curriculum? Must it possess any or all of these charac-teristics?" (Mukherjee, 1918)

In some senses Mukherjee was pointing to the fact that different universi-ties took on their identities largely as a result of the functions they chose to emphasize. On the one hand, they ranged from universities that were largely affiliating universities, granting degrees through the conduct of examina-tions. On the other hand, there were universities that were, in their own small way, trying to establish themselves as research universities, making way for the centrality of the Professoriate. While a healthy system of higher edu-cation will have room for different kinds of universities, Mukherjee was con-cerned with one question: Who should define the identity of a university?

Mukherjee's own starting point was a conception of a university as "A cor-poration of teachers and students, banded together in the pursuit of learning and for the expansion of the bounds of knowledge." Mukherjee was acutely aware that the historical, social and material conditions under which a uni-versity dedicated to these ideals could flourish were rarer than commonly supposed. Indeed, the bulk of Indian Universities were primarily dedicated to "certification", not the production of knowledge; and the curriculum was oriented towards servicing the state, or as a counterpoint, the reproduction of traditional forms of knowledge. In the debate that took place in India at the turn of the century, Mukherjee presciently identified a number of condi-tions that would have to exist for a liberal university to flourish. This short paper concentrates on three fundamental challenges for a liberal university in a context like India, but more generally. I conclude that this is a moment of precarious promise for the establishment for liberal universities in India.

ORGANIZATIONAL FORM/FINANCING

The conceit of the liberal university is the idea that it engages in the pur-suit of knowledge for "its own sake". What organizing and financing form would support such a university? Mukherjee very presciently understood that in some ways the university would have to be shielded from two diametri-cally opposite logics. On the one hand, it would have to be shielded from the bureaucratic impulses of the STATE; on the other hand, it would have to be shielded from being dominated by COMMERCIAL considerations, a calculus of returns on investment. This was particularly challenging in the context of a poor under-capitalized country, where a bulk of the financing of universities would likely come from the state. In such a context, the chal-lenge would be to design organizational forms that made the universities accountable, but did not impinge on their autonomy. Can a university be financed by a state, without succumbing to the imperatives of state power?

On this issue, the record of Indian universities is decidedly poor. At the turn of the century there was something of an elite compact, which tolerated the autonomy of universities. But this compact was very fragile and uneven, and by the 70s had become limited to a few elite institutions. The threat to universities came, in some instances, from direct politicization (the wholescale decimation of a university culture in West Bengal, the original site of Sir Ashutosh's hopes, being the prime example). But, more insidiously, it came from the logic of bureaucratization. In most state universities, the answer to the question: "Who gets to define the identity of the University?" was answered in one simple word: "The state." Mukherjee's hope that the state could finance universities and, yet, let the university community define the identity of the university both in terms of intellectual content and allocative decisions, largely came to nought. All the basic decisions of the university, what you can teach, how you can teach, who can teach, how much can you pay, how much can you charge, largely went out of the control of the universities. The imperatives that led to such state control were complex and need not detain us here. But suffice it to say that cumulative state control over the organizational form of the university impeded innovation and excellence. Indian universities were over-regulated and under-governed.

India is in a paradoxical situation where, on the one hand, there is a deep recognition of this fact. There is widespread acknowledgment that Universities need to be progressively given more autonomy. Several measures have pointed in the direction. A "graded autonomy" scheme has been introduced where universities will get a degree of autonomy depending on "ratings" carried out by a bureaucratic agency. At the extreme end is a scheme called Institutions of Eminence, which will free a select group of institutions from regulation altogether. The idea is to give a select group of institutions the freedom to define their own identities, set their own norms, subject to periodic reviews in terms of the progress they make in climbing up globally accepted ranking indicators. So, on the one hand, there is an acknowledgement that being "world-class" requires autonomy of action; unless a university is free to define its identity, it cannot attain excellence.

On the other hand, the quest for political control continues. Some of India's most influential public universities are the great sites of political contention. Many public universities at the regional level were often made subordinate to the ideological imperatives of the state. But, with the rise of populist/nationalist political parties, there is greater pressure on universities to serve the "national" cause. For example, this contention has taken an extreme form in one of India's most prestigious universities: Jawaharlal Nehru University. In some ways the university was always associated with being a bastion of "The Left". Whatever the truth of that contention may be, the University was accused of being "anti-national" with sedition charges being imposed on its student leaders. This is not just an isolated instance. The point was to send a message to all universities that unless they served the cause of authorized forms of nationalism, narrowly defined, their freedom would be curtailed. In some ways, universities have always served national projects, and forms of critical thinking that question nation state ideologies have often been suspect. But the recent rise in nationalist politics is putting universities under even greater threat, putting at risk the core freedoms of a university: the freedom to think.

So Mukherjee was far too sanguine that state funding could be made compatible with an organizational form that allowed functional autonomy to universities. But how does the private space fare in this context? Until very recently, the idea that private universities could create the free spaces required for learning and research had not really been tested in India. For one thing, no private university positioned itself as a major research university; most focussed on professional education. There were very few universities that focussed on the basic sciences and liberal arts. Most private universities were also subject to heavy regulatory control, including on curriculum and fees. But, most importantly, most private universities were closer to commercial enterprises, driven largely by revenue considerations. Most private colleges were oriented to professional education. In fact the early phase of private higher education in India was largely a product of the regulatory arbitrage. The state controlled the regulatory bodies that gave permission for colleges to be set up, and it requires considerable political manipulation to get permission to set up colleges. One striking manifestation of this was the fact that, according to one study, close to 80% of private colleges set up in India were set up by politicians of families with political connections. In short, the private higher education revolution in India was itself a product of an unholy nexus between state and capital — far from the insulation from state and capital that Mukherjee had dreamt of.

In the last few years there is beginning to emerge a new organizational form for a private university. This organizational form is relatively new to India. It is based on collective philanthropy. New universities like Ashoka and KREA are the nascent products of this organizational form. The collective philanthropy model has a few advantages. It ensures that the university is not an extension of the will of one or two proprietors. It ensures that governance processes in the university have to be relatively strong since attracting new donors requires credibility in process. In principle, such an organizational form should allow the university a degree of insulation from both the state, and immediate commercial considerations. But this is a very nascent revolution in India. Ashoka has demonstrated some early success with this model and has quickly gone on to become India's leading Liberal Arts university. But it is still an open question whether the cultural and political preconditions exist for such a model to acquire widespread currency. This model requires a widespread culture of relatively "dispassionate" philanthropy. There is a new generation of philanthropists — largely first-generation entrepreneurs, with strong experience of American universities — who are willing to go down this path. Given that the minimum scale of a viable research university in Indian requires at least $600 to $700 million in philanthropic commitments, it is not clear how many projects of this kind can take shape. This model is also still politically vulnerable in two respects. It requires regulatory clearances that still require "managing" the state; and it will require a state culture that does "pressure" capital and prevent it from funding liberal universities.

India is at the moment experiencing a tension. On the one hand, there is the prestige of the "liberal arts" model, as evidenced in the demand for admissions to top US schools; there is a desire to emulate the success of top global universities, and there is new Indian capital willing to take a bet on Indian Higher Education beyond professional schools. On the other hand, there is desire for regulatory control, formal or informal, the political pressures to enlist in the nationalist or other political projects, and the relatively small size of capital available. How will India navigate this tension? In all likelihood, there will be some room for innovation, since India has to cater to great demand. But India's full potential in the space of liberal Arts universities will still be hobbled. Ashutosh Mukherjee was right: a liberal university depends upon society providing organizational autonomy, between state and capital. We need to reflect on the conditions under which this autonomy can be taken for granted.

Just one more footnote on organizational form. In India much of the debate over university autonomy has meant "autonomy for the vice-chancellors". But what is the right combination of autonomy with accountability within a university remains a very unsettled question. India is still struggling to find an organizational form where the allocation of powers between the "professoriate" and "administration" is conducive to the overall aims of the university.

SOCIAL INCLUSION

We cannot take the organizational form that guarantees university autonomy for granted. But, in a poor country, marked by deep social and economic inequality, the "legitimacy" of elite universities is always open to question. The state was mindful of the social location of universities. A higher education system would be "tolerated" only in so far as it provided a means of social mobility and is not simply the site of the reproduction of social inequality. Arguably, this is an area of concern globally. Much of the "political" backlash against elite universities is fuelled by the sense that these are not socially inclusive spaces. Often this backlash is experienced simply through exit, a large majority of citizens do not think these are universities where their children belong. The role of universities reproducing rather than mitigating social distinction is a matter of global debate. Most universities recognize the importance of the issue. Affirmative action and diversity programs are designed to mitigate invidious forms of social exclusion that have marked universities. Yet it is hard to argue that universities, or the process to get to them, have been socially inclusive.

Ashutosh Mukherjee had raised this issue as well. Should a society worry about elitism of universities? He thought, quite rightly, that intellectual elitism was inescapable. But he was sanguine that universities did not have to worry about social elitism as much. The ultimate worth of the intellectual elitism would be redeemed by the fact that these institutions would produce graduates who would be exemplars in thinking about the public good. The university would become socially inclusive through the actions of its graduates and their impact on society. This view was extremely sanguine about the role of universities in creating just societies through the action of their graduates.

But societies do measure their universities on the scale of social inclusion. This was a truth that the democratic state in India recognized. Its answer was twofold. It introduced wide-ranging reservations for historically marginalized groups, where the aim of the universities was to mirror the social composition of society. This affirmative action has been the subject of great political contention. But this was also one of the reasons why there was political pressure to keep fees low. One of the criticisms public universities faced was precisely that they were unable to mobilize resources or signal the value of education by not pricing it right. The effects of these of these policies can be debated. They often ended up giving massive subsidies to the middle class as much as they enabled marginalized groups. But they signalled the fact that the university had to be positioned as a socially inclusive institution.

The dilemma for India is this. As the space for "private" education opens up, will the university remain a socially inclusive space? New universities like Ashoka are committed to social inclusion, through generous financial aid programs, with over 60% of students getting financial aid, and an outreach program that recognized social disadvantage. But there are three major challenges. First, the amount of philanthropic commitment and cross subsidy required to sustain a genuinely inclusive model is quite massive. Indeed there is anecdotal evidence that socially inclusive private universities do not do badly in reaching out to socially marginalized groups with incomes under five lakhs a year: conscious outreach and targeting can help. They also do well with privileged groups. But it is the lower middle they miss out on, where the signal a high price tag sends tends to socially deter these groups. If one were brutally honest about it, even a genuinely "needs blind" admission policy is sustainable only on the basis of prior inequality that is encoded into the admission and selection process. In a country like India a fully needs blind admissions policy would require foregoing almost 80 to 90% on the yield curve. Second, universities are built on the top of great inequality in school education and are yet expected to compensate for the inequality inscribed at the school level. The representation of the most marginalized groups in higher education is hobbled by the fact that the pipeline that funnel of applicants coming from the school system gets narrower the lower down the social or class order one goes. Third, and finally, there is the challenge of the university as social spaces. One of the challenges of elite universities is the fact that their culture is such that often students feel they don't belong there. Even if the university is financially inclusive, the challenge of creating a socially inclusive space. Imagine the challenges of creating a space where a first-generation Dalit student, whose parents are barely out of bonded labour, inhabiting the same space as a fifth-generation millionaire. Even in democratic societies, there is often a polite veil thrown over the fact that these spaces are difficult to create.

Higher education is about intellectual distinction. But the social legitimacy of universities is measured by their social inclusiveness. This social inclusiveness is a pedagogic necessity; it is a requirement of justice. But it is also a prudential political requirement. A university has to be a public trust in this respect: it has to be place where everyone potentially belongs. This is easier to announce than it is to credibly realize.

CURRICULUM

Even at the turn of the 20th century Mukherjee recognized that the liberal university's curriculum will aim to achieve some distinctive goals within a framework of overall excellence. But, as Mukherjee realized, in institutionalising the curriculum, there are tensions between these principles.

1. Breadth: The University must provide 21st century "Intellectual Literacy". What are the contours of 21st century Literacy that allows students to function in varied contexts?

2. Depth: The must be able to claim some credibility in a particular "discipline". At one level this demand is unexceptionable. But the "competence" requirement in each discipline is going up. Typically more and more majors require upping the number of courses required for the major. There is a tension emerging here between breadth and depth.

3. Diversity: The University should be a place where students find their intellectual identities; Students will have a diversity of abilities and temperaments. Each should be able to find their own measure. But does diversity of pathways pose obstacles to the signalling function of the University?

4. Choice and Boundary Crossing: The program structure must enable enough choice. For those students so inclined, there must be the possibility of crossing traditional disciplinary boundaries with credibility.

5. Core: Is there a "common foundation" to a liberal education? This is probably the greatest area of contention in curriculum construction. Broadly speaking there are three points of contention: What is a core stock of knowledge in the context of immense historical and social diversity? Should the core be a "substantive" core or a "methodological" one, organized around styles of thinking? How much of the curriculum should the core occupy?

6. Enablers: The imparting of enough core "skills" that are enabling conditions for all of the above. Initially this list included languages, writing, logical reasoning, but now includes extensive mathematics, programming etc. The biggest tension comes from the fact that serious mathematics is not just becoming part of 21st century literacy but a non-negotiable requirement for most majors.

7. Values: To what extent is the universities capable of imparting "values"? This was very much part of the project of liberal education, both in terms of substantive moral and civic values, but also a disposition to pursue higher values in general. What is the best way of thinking of university as being, to some degree, a site for the inculcation of values?

8. Research: the enchantment of university is not the transmission of knowledge, but the capability to "produce" knowledge, snatch snippets of intellectual order from a chaotic and complex world.

9. Contextual embeddedness. There is little point in disputing the fact that most elite universities take their cues from a global context of the production and dissemination of knowledge. But India in particular faces a peculiar challenge. It is relatively easy for elites to secede from their own contexts, and limit their scope for being meaningful change agents in their society. In India this tension is most apparent since most students from Indian elite universities find it difficult to function in contexts which require mastery of the vernacular. Indeed the suspicion of liberal arts as an "elite" project largely comes from its association with English, and the relative weakness of "vernacular" universities. What would it mean to produce graduates who could navigate the global and vernacular worlds with equal facility?

These curricular challenges are familiar to universities across the world. But in the Indian context the resolution of these tensions has been difficult for a number of reasons. The first is simply regulatory. The Indian regulators have been reluctant to allow four-year undergraduate degrees (with some exceptions). But India higher education will realize its potential only when it finds a creative way of harmonizing or at least mitigating some of these tensions.

India can be a propitious site for the creation of new dynamic liberal universities. It should aspire to be a global higher education hub. But it will first have to create first-rate exemplars of institutions that embody the organizational form, social legitimacy and curricular content of a liberal university.

REFERENCES

Mehta, Pratap Bhanu & Kapur, Devesh, Eds. (2017) *Navigating the Labyrinth: Perspectives on Indian Higher Education*, Orient Blackswan, Hyderabad.

Mukherjee, Ashutosh. (1918). Convocation Address, Delivered to the University of Mysore, published in *Dacca Review*, October.

PART III

· · · · · · · · · ·

The Future

CHAPTER 15

The Global and the Local: Constructing a Distinctive Role for Universities in Shaping the Future

David W. Leebron[1]

BACKGROUND OVERVIEW

The modern university traces its origin back to the founding of the University of Bologna in 1088. Universities quickly became the centres of scholarship and learning, and, while they grew significantly, they evolved slowly. In the early 19th century, the concept of the research university emerged in Germany. Universities became engines of technological progress. The German model was exported to the United States with the founding of Johns Hopkins in 1876, and both older (e.g., Ivy League) and newer universities (e.g., Rice and Carnegie Mellon) followed that model (Lucas, 1994; Britannica, 2019). This was accelerated further in the United States as the government relied on and funded universities for technological research for military purposes during the Second World War. Following the war, universities, most particularly Stanford, emerged as centres of entrepreneurial advancement and technological discovery for civilian purposes. Research funding expanded rapidly as the government launched major endeavours in space exploration, cancer and other health issues, and energy.

[1] I wish to express my thanks to Prof. Rebecca Richards-Kortum, Prof. Pedro Alvarez, Erica Ogwumike (Rice' 19) and Ryan Kirksey for their assistance.

The social role of universities also changed. Most universities, even highly renowned ones, remained fairly regional in most respects even into the second half of the 20th century, when they became more national and international. Over time, the universities evolved from being bastions of privileged students (white, male and wealthy) to being increasingly diverse engines of opportunity.

Universities remain complex in both organization and differentiation. They tend to be balkanized into schools and departments focused on historical disciplinary ideas. Centres and institutes are often created to overcome such balkanization and build interdisciplinary endeavours to address complex problems that require not only the knowledge and tools of a variety of disciplines, but new knowledge and tools that result from intellectual endeavors across disciplines.

The vast majority of institutions of higher education and research that are denominated as universities pursue a three-fold mission: teaching, research and service. (This is also true of many four-year colleges in the United States, although the balance among the missions differs significantly.) The nature of the missions varies a great deal, depending on the overarching institutional identity, its scope and reach. In the United States these institutions are either public (created under the auspices of a state, not federal, government) or private. The private institutions are either non-profit secular, sectarian (church affiliated) or for-profit (although the latter are generally not significant participants in research). According to a recent count, there are 4,298 institutions of higher education in the United States, of which 1,626 are public, 1,687 private nonprofit and 985 for-profit. Depending on their size, funding source, history, affiliations and location, the universities might conceptualize their mission primarily in local, state, national or other (e.g. religious) terms. There are 328 doctoral universities in the United States, of which 115 are categorized as R1 or "very high research" (The Carnegie Classification, 2019). The very high research universities range in size from 2,200 students (Cal Tech) to over 71,000 (Arizona State). Although the mix of such institutions varies greatly from one country to another, most universities fit in this broad categorization.

THE FORMS OF UNIVERSITY CONTRIBUTIONS TO SUSTAINABILITY

In light of this complexity, it is difficult to generalize about the role of universities in sustainability, as those roles vary according to the nature of the institution and are internally fragmented within universities. Generally, the contributions universities make to sustainability fall into five categories:

basic research, applied research and technology development, educational programs, the university's own sustainability practices, and service to external people or organizations that will benefit from assistance in one form or another. These categories of course overlap and blend into each other. Each of these modalities may be pursued with local priority, state priority or with a global perspective.

Sustainability at universities begins with their own university community. Residential universities are essentially small cities, providing the full range of services including housing, dining, transportation, police and healthcare. Universities are significant purchasers of a variety of inputs, including food, water and energy, and engage in substantial amounts of construction. And because universities want to apply insights gained from research in areas such as sustainability, they are constantly updating their practices to reflect knowledge and values. We see strong efforts by universities to reduce their carbon footprint, to build in environmentally friendly and sustainable ways, and to encourage behaviours that are less costly in environmental terms, such as recycling and limiting food waste. LEED certified buildings and environmentally friendly transportation (in both technology and community usage) have proliferated on American campuses, and administrative personnel help determine and guide practices that promote sustainability.

But the larger role of universities comes from their impact beyond their own campuses, whether in their own surrounding community or across the globe. Both urban and rural universities typically undertake both scientific and policy studies aimed to understand and benefit their immediate surrounding areas. At Rice University, for example, our professors have played a critical role in understanding the sustainability challenges of a coastal city, particularly one that regularly faces severe tropical storms (hurricanes). One effort is the university's Severe Storm Prediction, Education & Evacuation from Disasters (SSPEED) Center, which aims "to be recognized as the Gulf Coast's top university-based resource for research and education related to protection strategies for severe storm flooding and hurricanes-related surge" ("SSPEED Center: Vision," n.d.). Locally, it is often only universities that can research deeply into such problems as sustainability and help formulate solutions, as local entities rarely have such research capabilities.

THE INTERNATIONAL CHARACTER OF UNIVERSITIES

The impact of research universities extends well beyond their own communities. While major universities all claim a global or international role and perspective, in reality they remain overwhelmingly domestic institutions where international relationships are, for the most part, bilateral and transactional. They are significantly engaged in international trade of the services they produce through the mechanism of customers (students) travelling to the site of the enterprise to consume educational services. A few top research universities now enrol over 20% international students, and most are in the range of 10-20% of their undergraduate student bodies. (Institute for International Education, 2019.) The share of international graduate students, especially in STEM fields, is several times higher. There are a relatively minor number of foreign branch campuses of US universities and students studying at those campuses. Overall, it might be said that in terms of internationalization, higher education still resembles more of the 19th century model of transnational business rather than the 21st century global enterprise model.

However, a different story emerges if one looks not so much at the educational role as the research role. Faculty collaborations frequently span borders, although the vast majority of such international collaborations out of the United States are with researchers in other developed countries or in China. Deep collaborations with developing country universities and researchers, however, are rare. In addition, the faculties in the United States have a strongly international character. At Rice we looked at the fairly conservative measure that counts only faculty who received their first higher education degree (college BA or BS or similar) *outside* the United States. (Thus a student from another country who began his or her higher education at a US college would not count, even if they did so as a non-immigrant foreign student.) By that measure, about 31% of our faculty is international, and that is an important element of building the international research relationships and graduate student pipelines.

Equally important, the exchange of research information is global and frequently nearly instantaneous. Thus the exchange of ideas and results around much research, particularly fundamental research, has a strongly international quality. That has actually long been the case for universities. In the 19th century for example, the competition and intellectual exchange between the different schools of thought across national lines (particularly French, German and British) played a critical role in the successful development of the structural theory of organic chemistry (Hugill & Bachmann, 2005). Certainly the internationalization of science was a key part of the stunning developments in early 20th century quantum physics as well.

In sum, across the United States, we see a wide variety of international engagements, from professor-driven, two-person collaborations to the still quite limited establishment of foreign campuses or larger scale joint research enterprises. Each of these engagements reflects largely the structure of the home university, the benefits offered by particular foreign locations (hence a concentration in China, Singapore and wealthy Gulf states in the Middle

East), local demand and international accessibility. International teams in critical areas of sustainability (e.g. understanding climate change, creating sustainable technologies) are common. What remains comparatively rare are large scale and deeply rooted international research collaborations. Thus, while the international impact of research universities on sustainability is large, the primary channel for such impact is the effect of that research on similar challenges wherever they may be found.

UNIVERSITIES AND GRAND CHALLENGES

As universities are becoming increasingly international, they are also increasingly engaged in addressing concrete problems, often with funding from government research agencies, private foundations and individual charitable giving. A number of universities have explicitly decided in their strategic plans or other processes to identify "grand challenges" that they will focus on helping solve. In many instances, these challenges are locally formulated. For example, UCLA announced in 2013 the "Sustainable LA Grand Challenge" designed specifically to transition LA to a number of sustainable goals around water, energy and health. Its second selected grand challenge is "Depression", which it identifies as "a campus-wide initiative aimed at cutting the burden of depression in half by 2050" (Transforming Los Angeles, n.d.).

Such "Grand Challenges" vary in specificity and geography. At the University of Melbourne, for example, the three Grand Challenges are very broadly defined: Understanding our place and purpose; Fostering health and wellbeing; and Supporting sustainability and resilience (Research: The University of Melbourne, n.d.).

At Rice, our strategic plan didn't focus on grand challenges, but as part of our research aspirations it stated: "We should identify critical global challenges in areas such as health, education, cities of the future, and sustainability, energy and the environment to which Rice can make distinctive contributions, and work with partners locally and globally to achieve meaningful progress." Indiana University took a more specific approach that was also tailored to its role as the preeminent public university in the state of Indiana. Its three grand challenges chosen so far are: Precision Health Initiative, Prepared for Environmental Change, and Responding to the Addictions Crisis (Grand Challenges, n.d.).

A UCLA report in 2018 on "University-Led Grand Challenges" noted that "nearly 20 North American universities are leading Grand Challenge programs that are rallying research communities to contribute to solving a major societal challenge; attracting new investment and resources; demonstrating value of university research; and engaging students, partners, the

broader community, and the public" (Popwitz & Dorgelo, 2018). Its appendix identified 12 examples of university-led grand challenges, all aimed at
setting important research and education priorities that will address critical
problems. A plurality of such Grand Challenges appears to focus on a range
of sustainability issues, and a clear majority address sustainability and health/
medicine.

These programs vary greatly in terms of mission, scope, specificity, funding and partnerships. Not surprisingly, these grand challenges tend to focus
largely on local jurisdictional benefits and to some extent the benefits to
the specific mission of the university, such as educating its students. What
drives them in many respects is a sense of high ambition, a desire to cap-
ture increasingly programmatic private funders, and seizing on governmental
funding opportunities, both local and national. For the most part, they seem
aimed at coalescing and coordinating existing strengths and programs and
supplementing them with additional resources and other forms of university
support. In sum, they appear to be more about prioritizing and coordinating
than truly doing things differently (although some might observe that for
many universities, prioritizing and coordinating is in fact doing things differently). Thus, it's not clear that the identification of the grand challenges
implements a different role for the university in addressing the large scale
problems faced both locally and globally.

UNIVERSITIES AND SUSTAINABILITY IN GLOBAL SCALE

Universities have rightly become seen as substantial contributors both to
local economies and to the solution of national and local problems. But,
despite the proclamation of grand challenges, universities are notoriously
bad at formulating and sustaining highly focused efforts, and there are multiple reasons for this that are deeply embedded in university culture, historic
practice and values. In addition, their track record in working together to
create global approaches is limited. (Huge exceptions include the CERN
effort in particle physics and large scale telescopes.) Of course, virtually every
solution to a domestic problem has benefits for similar problems elsewhere
around the globe, even if significantly affected by local conditions. Thus
there is a trickle-down (or perhaps more appropriately, trickle-out) approach
for universities to achieve global impact.

A look at available information on higher education's role in achieving
the UN's sustainability goals is fairly discouraging. For example, the Higher
Education Sustainability Initiative (HESI) contains little that is concrete,
convincing or impactful. The emphasis seems to be primarily on membership
and conferences. Though HESI claims that it "provides higher education
institutions with a unique interface between higher education, science, and

policy making," the evidence of that seems limited (Higher Education, n.d.). However, one example presented at the HESI conference appears to be a good example of a collaborative international education initiative aimed at making contributions to sustainability, namely the Geneva Tsinghua Initiative for Sustainable Development Goals. This program appears to integrate efforts across institutions from a developed and developing country and build deep relationships among students and others. The educational approach is also broadly integrative across methodologies and purposes, from traditional educational environments to online modalities to entrepreneurial and maker spaces.

A number of universities explicitly aim to develop exportable or scalable technologies to address sustainability and other challenges. But, in fact, such technologies often turn out not to be exportable to lower resource environments, at least in the near term, for a variety of reasons. These include cost, maintenance issues, lack of local materials and manufacturing capacity, inadequate educational training and capacity, lack of cultural fit and other unanticipated collateral costs and obstacles. To give just one example, a low-cost diagnostic test that took several days to produce results might not work in an environment in which a patient had to travel for a day to a clinic or hospital from her village, and couldn't afford to wait or to make another trip.

Many of the grand challenges involve health issues, such as curing cancer. And while ultimately the knowledge gained will benefit communities around the globe, the process will be slow and often require years of adjustment to local resources and conditions. The spread of solutions can be further hampered by the creation of intellectual property, the deployment of which is determined largely on the basis of financial return.

INTEGRATIVE EXAMPLES FROM THE RICE EXPERIENCE

There are two efforts led by Rice University faculty that suggest comprehensive solutions-oriented research approaches that span institutions are possible, and some of the essential elements for success.

Neo-natal Care: Nest 360

In 2018 the MacArthur Foundation set about the process of identifying the recipient of a $100 million one-time grant for a project that "promises real and measurable progress in solving a critical problem of our time." In the words of the foundation, the essential requirement was that "the proposal describe the urgent problem worth solving, and [that] the solution have a transformative impact." The solution was required to be evidence-based, feasible and durable (100 and Change, n.d.).

In many ways, the results of this competition revealed that universities were not, at least in the judgment of the foundation judges and board, the entities best positioned to address such problems at scale and with urgency. Only one of the eight semi-finalists was primarily a university entity. Four of the projects addressed human health issues directly, one food supply and health, two education broadly and one social welfare programs.

The only university-affiliated semi-finalist was a project organized by Rice professor Rebecca Richards-Kortum and others, called Nest 360, to solve the challenge of over one million babies who die each year in sub-Saharan Africa largely from preventable causes. While technologies existed in developed countries to prevent these deaths, such technologies were not sustainable in developing countries both because the cost was too high and they could not be manufactured or maintained locally. Much equipment ended up in "equipment graveyards" as a result. Rice 360 (the entity within Rice out of which this project emerged) integrated a set of 17 technologies that if implemented as part of a neo-natal suite developed by Rice would prevent at least half of such deaths. But technology development, which universities can excel at, was only part of the problem. Rice 360 identified four "gaps" and corresponding work streams: innovation (including manufacturing), education, implementation and market shaping that would generate demand and create a distribution channel.

In short, a sustainable solution in health care required the creation of a complete ecosystem that addressed all aspects of a solution and provided an adequate feedback loop for the continuous evolution of the solution. To address this, Rice 360 expanded a complex set of collaborations aimed to bring expertise to diverse tasks and build local capacity where needed. The partners included two key local university partners, namely the University of Malawi College of Medicine and Malawi Polytechnic, as well as a local hospital, the Queen Elizabeth Central Hospital in Blantyre. One specialized international higher education partner, the London School of Hygiene and Tropical Medicine, was also part of the consortium, as was a domestic partner chosen to bring business and logistics expertise, the Kellogg School of Management at Northwestern University. Finally, since the project involved the production of physical equipment, a design and manufacturing company was added, 3rd Stone Design, which emphasized the integration of "user needs, environmental constraints, technological capabilities and economic realities to create convincing solutions to difficult problems." (The presentation to the MacArthur Foundation judges can be seen online [Macfound, 2017]). As the project has expanded beyond Malawi, additional partners have been added, including the Ifakara Health Institute in Tanzania, the Dar es Salaam Institute of Technology, the University of Lagos, the University of Ibadan and Kenya Pediatrics Association.

Although Rice 360 did not win the competition for the $100 million, they received a $15 million award from the MacArthur Foundation that has launched them on the path of achieving grants from multiple foundations that will enable them to complete the first phase of the project in Africa.

This project exemplifies the impact that universities can have with the right partners in addressing sustainability challenges. The first element is a core group of researchers and staff driven to have an impact on the world. While the university provided smaller, strategic support at early phases, the bulk of the funding has been external. Careful development of long-term relationships with universities and other institutions on the ground was critical, as were partnerships with universities in developed countries that could provide critical expertise. In sum, the entire chain from innovation to implementation needed to be enabled and sustained by personal, institutional and funding commitments.

Solving Global Water Problems: NEWT

A somewhat more traditional example of large scale sustainability impact emerging out of Rice is Nanotechnology Enabled Water Treatment (NEWT). In the words of its website, "NEWT is an interdisciplinary, multi-institution nanosystems-engineering research center (headquartered at Rice University) whose goal is to facilitate access to clean water almost anywhere in the world by developing affordable and efficient modular water treatment systems that are easy to deploy, and that can tap unconventional sources to provide humanitarian water or emergency response" (NEWT, n.d.).

This is an effort led by Rice Professor Pedro Alvarez in collaboration with researchers at a diverse set of four universities: Rice, Yale, Arizona State and University of Texas El Paso. That collaboration of four universities enabled NEWT to receive an initial five year renewable NSF grant of $18.5 million to establish an Engineering Research Center "to develop compact, mobile, off-grid water-treatment systems" (Boyd, 2015). In addition, universities in China and Brazil have also been engaged, in part to provide on-the-ground expertise, testing and partnership in locations in need of such technology. In both cases, available national resources drove part of these collaborative efforts. NEWT leaders also recognized that sustainable success would depend on industry partners, and engaged nearly 20 such partners across the potential value chain from manufacturers of materials and equipment to service providers and end users.

Such collaboration was helped by a limited emphasis on the exploitation of intellectual property, but a tough-minded approach to practicality. Alvarez subscribes to an adapted version of the NABC value creation method suggested by Curt Carlson, a leading thinker on innovation (Denning, 2015):

starting with the identification of **N**eeds, adopt an **A**pproach that is appropriate and distinctive, and consider the **B**enefits in relation to the project's costs, as well as the **C**ompetition and alternative solutions. While the research being undertaken is of a kind universities typically engage in, the approach, mission and partnerships help assure broader and larger scale implementation.

These examples still stand in many ways as exceptions. Effective comprehensive partnerships that can address sustainability issues across the developmental spectrum are few. As Inside Higher Education reported just last fall, "it is striking that partnerships between the poorest nations and the world's research elite form a very small slice of international collaboration" (Baker, 2018). According to the inside higher education analysis, among the top 10 universities "less than 3% of cross-border research featured a partner from nations categorized ... as the world's least developed. At four of the universities, the share was lower than 1%" (Baker, 2018). The largest share of such collaboration was medical research.

Equally, one can look at student flows and see similar shortcomings, although not quite as bad. According to the IIE's Open Door studies, over 60% of American students studying abroad do so in developed countries.

The barriers to the kinds of collaboration that might make deeper and faster progress on global sustainability questions are entrenched. First and foremost are the nationally directed funding sources. At least in the United States, the major research funding agencies have limited willingness to fund efforts outside the country. USAID has a good track record of working closely with universities to support efforts with impact on developing countries, but some reports suggest that willingness has been reduced in recent years. One example is the Higher Education Solutions Network, "a partnership between USAID and seven top universities" aimed at fostering innovation to address development challenges (H.E.S.N, 2018). Similarly, the Partnerships for Enhanced Engagement in Research (PEER) help foster partnerships between developing country scientists and those in the United States. International funding agencies, such as the World Bank, seem to play a very limited role in supporting the contributions that universities could make to large scale sustainability efforts.

One notable US university-based effort that represents at least a partial integration of researchers from around the globe to address a congeries of sustainability issues is the Global Resilience Research Network, organized by the Global Resilience Institute at Northeastern University. The GRRN "is a membership network of leading universities, institutes, non-profit organizations, and companies engaged in resilience research that informs the development of novel tools and applications". (Global Resilience Institute, n.d.) Its membership includes entities from every continent, although it is largely focused on the developed world and the Caribbean. The website, however, provides little clear indication of activity other than an annual summit and some facilitation of collaborative research.

SOME CONCLUSIONS

As Einstein famously said and is so frequently quoted: "The world that we have made as a result of the level of thinking we have done thus far creates problems that we cannot solve at the same level as the level we created them." One might argue there is a corollary to this quote: "The world that we have made with the institutional structures we have had thus far creates problems we cannot solve with the same institutional structures that created them."

On the optimistic side, we have already seen a major change in how universities contribute to fundamental solutions. There is greater production of intellectual property and greater collaboration with industry. There are more programs, institutes and focused collaborative research endeavours that aim to solve identifiable problems. More of our research enterprise is driven by increasingly massive amounts of data. Collaborations across universities are commonplace.

What are the special strengths universities bring, and what are the weaknesses, as we seek sustainability solutions that are both local and global? Compared to the private sector, universities are mission driven to achieve human welfare even when that doesn't translate into monetary return. They are good at developing fundamental knowledge and application strategies, and at their best able to use a range of available talent that includes undergraduate and graduate students as well as post-doctoral researchers, talented administrators and brilliant professors. Universities are far better positioned than most actors to integrate knowledge across disciplines, enabling them to simultaneously address, for example, engineering questions and cultural barriers to adopting solutions.

But, for the most part, universities are not good at focusing on a few projects or delivering fully integrated solutions to problems. The effective application of knowledge, and integrating knowledge into practical frameworks, is typically not their strength. Despite claiming global perspectives, a variety of pressures drives them to more locally oriented projects. Perhaps most frustratingly, problems are urgent and solutions are not. Universities tend to be too slow, and other actors often attempt to be too fast.

Large, globally oriented foundations are playing an increasing role in funding the solution of grand challenges that are not necessarily centred in the developed countries. In the US these mega-foundations include Gates (currently $50 billion), Ford ($12 billion) and MacArthur ($7 billion). On the other hand, foundations have a tendency to want to invest only for limited periods rather than the long run required, as well as to fund at wholly inadequate levels the infrastructure (overhead) required to make the project funding approach truly sustainable. Nonetheless, the emergence of major foundations explicitly committed to "strategic philanthropy" to address major challenges, including sustainability issues, is changing the landscape of what is possible. These foundations increasingly have both the resources and organizational expertise to help motivate and coordinate critical actors.

Universities on their own are generally not in a position on their own to discover, design and implement large scale sustainable solutions to major problems. Here are some practices and solutions that could enhance both the role of universities and their effectiveness:

1. Universities in developed countries must partner in long term, sustainable and respectful ways with universities and other partners in developing countries.

2. These partnerships must be funded in a sustainable way that doesn't put the burden on low income developing country partners.

3. A key part of these partnerships must consist in building capacity in developing country institutions and people.

4. All university partners must be involved in all aspects of the relationship, and exchanges and other aspects must be mutual, including opportunities for shared experiences and cultural immersion.

5. Ownership, learning and decision processes must be shared, and especially located in the country where challenges are being addressed. Planning and implementation must take into account local cultures and governance.

6. Processes must involve all necessary disciplines and processes must provide for the engagement of those disciplines from planning through execution.

7. Partners must be identified and engaged across the entire learn-advocate-design-build-distribute-manage-maintain-evaluate ecosystem. Such partners should virtually always include, along with universities, non-profit enterprises, for-profit businesses and responsible government entities at the appropriate levels.

Building effective partnerships characterized by trust and a shared mission is challenging, especially since 1) typically the effort will be only a part, and often a small part, of each partner's mission and 2) each partner's and individual's mission and incentives will be different. This will affect views on everything from how learning should take place to which tools will be viewed as most effective (i.e., "if all you have is a hammer, every problem looks like a nail") to the time horizons that are employed.

But as we have seen at Rice and elsewhere, new approaches to building partnerships, designed for deep and sustained collaboration and impact, can leverage the strengths of universities to truly address the world's most salient issues of sustainability and health.

REFERENCES

100 & Change. (n.d.). Online: https://www.macfound.org/programs/100change/strategy/

Baker, S. (2018). "Study Finds Limited Collaboration Between Research Elites and Developing Nations". *Inside Higher Ed*, 21 September. Online: https://www.insidehighered.com/news/2018/09/21/study-finds-limited-collaboration-between-research-elites-and-developing-nations\

Boyd, J. (2015). Rice, ASU, Yale, UTEP Win NSF Engineering Research Center. Online: https://news.rice.edu/2015/08/10/rice-asu-yale-utep-win-nsf-engineering-research-center/

Britannica, T. E. of E. (2019). "Johns Hopkins University". Online: https://www.britannica.com/topic/Johns-Hopkins-University

Denning, S. (2015). "How to Create an Innovative Culture: The Extraordinary Case of SRI", *Forbes Magazine*, 30 November.

Global Resilience Institute. (n.d.). Online: https://globalresilience.northeastern.edu/

Grand Challenges. (n.d.). Online: https://grandchallenges.iu.edu/

H.E.S.N. (2018). Higher Education Solutions Network (HESN): U.S. Global Development Lab. Online: https://www.usaid.gov/hesn

Higher Education Sustainability Initiative. (n.d.). Online: https://sustainabledevelopment.un.org/sdinaction/hesi

Hugill, P. J. & Bachmann, V. (2005). The Route to the Techno-Industrial World Economy and the Transfer of German Organic Chemistry to America Before, During, and Immediately After World War I. *Comparative Technology Transfer and Society*, 3 (2), pp. 158–186

Institute for International Education (2019). "Open Doors Report on International Education Exchange 2018". Also: https://www.iie.org/en/Research-and-Insights/Open-Doors/Data

Lucas, Christopher J. (1994). American Higher Education: A History.

Macfound. (2017). Online: https://www.macfound.org/press/semifinalist-profile/rice-360-institute-global-health-rice-university/

Moody, J. (2019). *A Guide to the Changing Number of U.S. Universities*. Online: https://www.usnews.com/education/best-colleges/articles/2019-02-15/how-many-universities-are-in-the-us-and-why-that-number-is-changing

NEWT. (n.d.). Online: http://www.newtcenter.org/

Popowitz, M. & Dorgelo, C. (2018). Report on University-Led Grand Challenges. UCLA: Grand Challenges. Online: https://escholarship.org/uc/item/46f121cr

Research: The University of Melbourne. (n.d.). Online: https://about.unimelb.edu.au/strategy/growing-esteem/research

SSPEED Center: VISION. (n.d.). Online: https://www.sspeed.rice.edu/vision

The Carnegie Classification of Institutions of Higher Education. (2019). 24 May. Online: http://carnegieclassifications.iu.edu/

Transforming Los Angeles Through Cutting Edge research. (n.d.). Online: https:// grandchallenges.ucla.edu/sustainable-la/

CHAPTER 16

Traditional universities: challenges and opportunities

Joël Mesot[1]

INTRODUCTION

What is at stake?

Never before has there been such a huge choice of providers of higher education as today. As this sector grows in reach and impact, it is also becoming more international. OECD data show that the member countries host more than 3.5 million international students; 6% of all students in tertiary education in OECD countries are international, and the number rises to 12% for masters and 27% for PhDs. (OECD, 2019) In response to this development, a few years ago *Times Higher Education* introduced a special ranking of the world's most international universities. The latest survey shows Switzerland, Hong Kong, Singapore and the UK as being home to the 10 most international universities in the world (*Times Higher Education*, 2019) The ranking is based on four groups of scores: international students, international staff, international co-authorship and international reputation metrics.

With globalization acting as one of the main drivers of economic growth, higher education has become a global affair, setting in motion a process of differentiation and the emergence of a plethora of new players. What does this all mean for a traditional university such as ETH Zurich in its 165th year

---

[1] The author would like to thank his Public Affairs advisor, Roman Klingler, for his support in writing the text.

of existence? The university's role is to prepare the next generation of engineers, scientists and leaders, and to shape the world through basic research and forward-looking education. So, what is at stake when not only competition with peer universities is fierce, but large corporations compete for talent and new educational providers challenge the business model of traditional universities? What needs to be done in order to marry change with tradition, and develop the university in a sustainable way? In what follows, we address these questions and provide some answers.

GLOBAL TRENDS AFFECTING HIGHER EDUCATION

End extreme poverty worldwide. Significantly reduce marine pollution and take action to combat climate change and its impacts. These are just some of the 17 demands the United Nations has set out in its Sustainable Development Goals (SDGs) that all member states adopted in 2015 within the framework of the 2030 Agenda. All countries are therefore called upon to come together to solve the pressing challenges of the world and commit themselves to sustainable development. Universities have a special role and responsibility in this global endeavour. The difficulty is to achieve this whilst navigating the complexities of the Fourth Industrial Revolution, as described by the WEF (World Economic Forum, 2017). How universities shape the way talent thrives is a key driver in the transition to a new work environment dictated by the scale and pace of technological innovation.

The race for technological supremacy and Asia's ambitions

Fueled by advances in robotics, data science, artificial intelligence and life sciences, we are witnessing a global race for technological supremacy. There are two main protagonists in this race — the US and China — while Europe tries to keep pace with the massive investments on either side of the Pacific. According to OECD data, China spent US$443 billion on research and development in 2017, second only to the US, with $484 billion. China produces more scientific publications than any other country, and in the next decade is likely to rank top for citations. (*The Lancet*, 2019). A similar development can be expected for patents. This shift in scientific and technological prowess goes along with Asia's ascent as an economic powerhouse. Asia-Pacific countries' share of global GDP was close to 43% in 2018, compared with 15% for the US and 16% for the European Union (*Die Volkswirtschaft*, 2019).

For many observers, the conclusion is crystal clear: if the 19th century was the zenith for Britain and the 20th century for America, the 21st century will belong to Asia. The confidence among Asian leaders is epitomized by intellectuals such as Kishore Mahbubani. In his book *Has the West lost it?*, Singapore's former ambassador to the UN not only predicts the inevitable growth of Asia's dominance, but sees this shift in geopolitical power as a natural development towards historical normality: "Viewed against the backdrop of the past 1,800 years, the recent period of Western relative over-performance against other civilizations is a major historical aberration. All such aberrations come to a natural end, and that is happening now". (Mahbubani, 2018).

The US is struggling to respond to this world-changing challenge. Unlike China, where the central government is pushing the implementation of AI technology, America's efforts seem fragmented and decentralized. In the words of Professor Amy Webb, a specialist in strategic foresight at the NYU Stern School of Business, "China is the OPEC of data. In an authoritarian society, every human and social interaction feeds a vast pool of structured data for machines to ingest" (*Washington Post*, 2018). Meanwhile, Europe tries not to be outstripped by the two dominant regions and is raising its financial bid with a total investment of €100 billion in the new Research Framework Programme "Horizon Europe", which will run between 2021 and 2027.

Tech giants push into basic research and compete for talent

Competition for technological leadership is not only between countries and continents: the digital era has also seen the rise of so-called "superstar" companies, with inevitable consequences for universities. Four out of the five US corporates with the biggest market capitalization are tech companies (Microsoft, Amazon, Alphabet and Apple). The value of three of these economic behemoths — Apple, Amazon and Microsoft — has at times hit the one-trillion-dollar mark. Tencent and the Alibaba Group are the two most capitalized companies in China. By comparison, Europe's big five comprises traditional industries (Nestlé, Shell, Roche, Novartis and Anheuser-Busch InBev). Europe's biggest software company — SAP — is not even in the top five.

Alibaba, Amazon, Apple, Baidu, Facebook, Google, IBM, Microsoft and Tencent are a group of nine tech giants that are instrumental in the development of AI (Webb, 2019). While the US government has largely outsourced basic research to the commercial sector, China's AI push is part of a coordinated attempt to create a new world order, argues Webb. These tech companies are so financially strong that they can invest billions in research and increasingly compete with universities for top talents. This is not only happening in the AI domain, but can also be seen in Google's secretive Calico project. Launched in 2013, this biotech company is trying to find the causes of ageing — a dream of many Silicon Valley billionaires. There is not much information available about the scientific activities of Calico, but the San Francisco based company seems to be generously funded, with $1.5 billion in the bank. Calico's Chief Scientific Officer, David Botstein, has described it as "a Bell Labs working on fundamental questions, with the best people, the best technology, and the most money". (MIT Technology Review, 2018).

THE DIGITAL TSUNAMI AND THE UNIVERSITIES

Towards a more personalized education

The advent of massive open online courses, or MOOCs, and other disruptors in higher education has led some observers to proclaim the end of the traditional university altogether. A decade on, this scenario has clearly not materialized, but technology — and particularly the potential of AI in education — will undoubtedly disrupt our concepts of knowledge acquisition and transfer. The dawn of the Fourth Industrial Revolution and the global trends described previously have encouraged a new set of societal expectations. Explicit knowledge will no longer suffice to prepare students for an ever-changing career path. Learning sciences have made a strong case that explicit knowledge needs to be combined with implicit knowledge in order to deliver the best educational outcomes. Implicit knowledge, as opposed to its explicit sibling, is hard to codify and is transferred most efficiently through experience-based learning.

These two forms of knowledge are, for example, at the core of an innovation project which ETH offers to students in mechanical and process engineering. The semester program, which is compulsory for all second-semester students in mechanical engineering, fosters critical thinking and is problem-oriented. Rather than acquiring knowledge about mechatronic relationships passively, students gather that knowledge on their own by working in small project teams. For support, the students can turn to coaches from more advanced semesters, who have enhanced their skills and experience in a tailored coaching course.

As our societies evolve and the educational functions of a university change, the need for a systematic and scientific way to look at learning grows too. At ETH Zurich, we are therefore investing in this field by launching the "Future Learning Initiative". This initiative aims to carry out interdisciplinary research on learning, and translate the basic research to build and test interventions and applications for deep learning at ETH. The initiative will see the establishment of new professorships, as well as projects that will not only tap into the potential of technology for learning, but at the same time reflect the role of humanities and ethical aspects in the education of engineers and natural scientists.

Fledgling universities and new kids on the block

Part of the impressive Asian story relates to the rise of relatively young universities. A number of fledgling institutions in China and other parts of South-east Asia have been built from scratch in recent years and have followed a fast track to academic and scientific achievement. The Southern University of Science and Technology (SUSTech) near Shenzhen, established only in 2009, is typical of this new brand of rapidly expanding institutions. Bolstered with important financial resources and vast state support, they are investing in the development of their campus and are recruiting scientists from all over the world with tempting offers.

Furthermore, tradition-rich educational vessels see themselves challenged by disruptive speedboats that come up with specially tailored offers. Minerva is a case in point. It claims to offer a reinvented university experience of small, online seminars delivered through a unique digital learning platform, combined with residential experiences across the world. The company does not hide its ambitions, openly declaring that it wants to become the world's leading university. To achieve this, Minerva has dissected traditional academic institutions structurally and pedagogically in order to identify their strengths and weaknesses. Putting the students at the centre, Minerva shifts the learning paradigm from imparting past and present knowledge to developing lifelong skills. These are not just the skills required for the jobs of today; the emphasis is on how to learn and adapt throughout life, so students can be ready for the jobs of tomorrow.

As learning technologies progress hand in hand with the changing needs of a new global workforce, universities will have no choice but to embrace this paradigm shift and adapt to cover a broader range of educational imperatives — or face competition in this field from new players attempting to fill this gap. From the delivery of knowledge to the facilitation of learning, more and more competitors are vying for space in an untapped educational niche. These factors, along with the advancement of digital learning, reflect the changing purpose of education, and by extension the role of universities in a societal context.

We have mentioned only one disruptor here, but there are many more — 2U and Khan Academy, Singularity University or Ecole 42 to name a few — and while their rapid rise has not spelled the end of traditional universities, which tend to have a much wider mandate in research, education and tech transfer, they are not going away either.

As the custodians of traditional academic institutions, we should take note: the persistence of these new kids on the block highlights the disruption taking place in post-secondary education and suggests that the university of the future will not look like the academic institutions of today. As Richard

DeMillo from the Center for 21st Century Universities at Georgia Tech puts it: "We need to rethink the nature of the contract between society and its universities." (DeMillo, 2015).

THE UNIVERSITY IN ITS CULTURAL AND NATIONAL CONTEXT

The weight of history — Humboldt's legacy

Every university has its own "genius" and history that in some cases extends back to medieval times when the first universities on European territory were founded. No matter how long this history is, the origins of a university transcend into the present and the future of an institution. Just as scientific advances are built on previous discoveries, traditional institutions benefit from the experiences of their predecessors. Many universities — including ETH — are imbued with the educational ideal of Wilhelm von Humboldt, the founder of Humboldt University of Berlin. His principles of academic freedom, the unity of research and education, and his holistic approach to education can still serve us well as guidelines. ETH Zurich's starting point is intrinsically tied to the advent of the modern federal state in Switzerland and its economic ambitions for development and industrialization in the mid-19th century.

When discussions started in the first half of the 19th century about the establishment of a national (federal) university in Switzerland, several cantons competed for the coveted status. Zurich was one of several possible locations, and the political compromise of the time was then to create a national school of engineering and natural sciences instead of a fully-fledged university. The institution's mission was set down in a special law of 1854 on the establishment of a Federal Polytechnical School, as ETH was called at the time: "The task of the polytechnic school is to train technicians 1) for road, railway, hydraulic and bridge construction, 2) for industrial mechanics, 3) for industrial chemistry, always taking into account the specific needs of Switzerland, theoretically, and as far as possible also practically." (Schweizerisches Bundesarchiv, 2019). In short: the new school was meant to train experts to build the necessary infrastructure for industrialization.

A hundred years later, Swiss politicians made another farsighted decision when the Ecole Polytechnique de Lausanne (EPUL) became the Ecole Polytechnic Federal de Lausanne, thus gaining the same federal status as ETH Zurich. The foundation of EPFL in 1969 as a Swiss Federal Institute of Technology paved the way for EPFL's outstanding development over the last decades to become one of the top technical universities worldwide.

Embedded in the ETH Domain

Our two leading universities — ETH Zurich and EPFL — are embedded in a national framework of scientific excellence, along with four research institutes: the Paul Scherrer Institute (PSI) where the Swiss large-scale user facilities are located, the Swiss Federal Institute for Forest, Snow and Landscape Research (WSL), the Swiss Federal Laboratories for Materials Science and Technology (Empa), and the Swiss Federal Institute of Aquatic Science and Technology (Eawag). All these institutions make up what is known as the ETH Domain, under the auspices of the State Secretariat for Education, Research and Innovation (SERI).

Every four years, the Swiss Parliament deliberates and sets the parameters for the country's education and research area. The Dispatch on the Promotion of Education, Research and Innovation (ERI Dispatch) encompasses vocational training, the Swiss National Science Foundation and universities (including universities for applied sciences, cantonal universities and the ETH Domain), and provides funding for the next four-year period.

It is within this framework — and based on the Federal Act on the Federal Institutes of Technology — that ETH Zurich and EPFL are free to set priorities and define their respective strategies. Although the institutions of the ETH Domain are independent of one another and are competitors in the global arena of higher education and research, they cooperate in a number of areas in the national interest. Some examples: both ETH and EPFL operate the Swiss Data Science Center and provide industry and other Swiss universities with access to expertise and infrastructure. Furthermore, the two work closely together in cyber security; 28 of ETH Zurich's professors conduct their research within one of the four research institutes of the ETH Domain. Since all institutions of the domain belong to the same legal structure, collaborations can be easily set up. This situation provides a strong competitive edge, around which future strategies of the ETH Domain must be developed.

Quadruple mission — education, research, tech transfer and dialogue with society

Since its inception some 164 years ago, ETH Zurich's core mission has not fundamentally changed. Its mandate is still to educate the next generation of engineers, architects and natural scientists. As a largely publicly financed university, social equality in the access to university is an important issue. As opposed to other (private) universities that require an entrance test, ETH is open to every prospective student holding a Matura, the Swiss secondary-school diploma. At the end of their first year of university, however, students have to pass a demanding test in order to continue their studies. Both systems have their advantages and drawbacks. I am convinced, though, that a test after one year is a fairer solution than raising the barrier right at the beginning, as it allows students at least one year to adapt to their new environment.

While research also goes on in industry, both basic and applied research remain key to innovation and the country's economic development. Whereas research results are persuasive vehicles for communication, it is far more challenging to persuade politicians to support basic research because of its uncertain nature and putative commercial use. Thirdly, knowledge and technology transfer belong to the core tasks of a university, a fact that has become even more important in recent years as politicians realize that technological progress is an essential condition for the nation's future economic prosperity.

A final dimension must be added to ETH's core mission: the dialogue with society. This basically serves three purposes: firstly, to explain to politicians and the Swiss taxpayer how the public funds are being used, and for what purpose. Secondly, an ongoing dialogue with various stakeholders prevents ETH from losing touch with social reality. And thirdly, it is imperative that universities play an active role in the social discourse over the introduction of new technologies and the multiple ethical questions change brings about. The crucial discussion on AI, for example, cannot be left to the tech giants and other interest groups. To address this issue on a more neutral footing, ideas have been put forward for an international hub for AI research linked to the UN, in which Switzerland could play an important role (Fischer & Wenger, 2012).

INSTITUTIONAL CHALLENGES

Stormy times for the university

Increased media coverage and a change in perception of hierarchies and institutional power among the younger generation are two reasons why personal conflicts in academia have become more public in recent years. ETH is no exception here. The institution has had to deal with several cases of misconduct and abuse of power by professors in their relationships with (doctoral) students (ETH Zurich, 2019) in its recent past. As conflicts in a highly competitive environment such as a university can never be completely avoided, the cases have brought to light not just individual misconduct, but also structural weaknesses. Allegations of bullying have not only led to intractable confrontation between the parties involved, but have unleashed tremors that have shaken the institution to its foundations. The extensive media coverage, fuelled by incessant leaks of confidential information, has tarnished the university's reputation.

For the first time in its history, ETH has approached the ETH Board for permission to terminate the employment relationship with one of its professors. As challenging as a crisis can be, it also presents an opportunity to become a better institution. The major lessons of these upheavals fall into three categories: prevention, leadership and management of conflict situations.

Strengthen leadership — reduce structural dependencies

Prevention starts with the selection process for people who join ETH. Leadership skills will be given more consideration when hiring new personnel. We have adapted the appointment criteria for new professors: now their leadership skills are being assessed, as well as their excellence in research and teaching, both of which are, of course, still crucial.

ETH will renew its commitments to diversity and inclusion, and ensure these topics feature prominently in the leadership criteria. Students, administrators, faculty and academic staff should reflect this commitment at all levels. Embedding this at the institutional level will demonstrate that diversity, fairness and inclusiveness are an integral part of our vision for the university.

Special induction programs for new professors, as well as for doctoral students, will address the expectations and values associated with a good working relationship. Leadership skills will be strengthened through coaching programs tailored to participants' specific needs. Furthermore, ETH will introduce multiple supervision for all doctoral students by 2020, along with a set of other measures to reduce the dependent relationship between professors and doctoral students. That said, it is important to note that completing a doctorate at ETH remains a challenging task, with no guarantee of success.

Conflict management — fair and swift processes

The problems explained earlier have also shed some light on processes and structures that need to be improved in order to prevent further escalation of personal conflicts. One such weakness was the fact that the conflicts were not addressed early enough. The number of ombudspersons has already been increased, reports of sexual harassment and inappropriate behaviour will be dealt in future by a specialized reporting office within the HR department and also through an external independent office. The process for dealing with complaints will be streamlined to ensure that all reports are addressed and if possible resolved within six months. Last but not least, ETH's leadership continues to raise awareness about respectful ways of interaction. A culture of "speaking up" when things go wrong, must be further developed. All this will require time, however.

Remaining open to the world

Switzerland's success story in terms of competitiveness and innovation prowess is regularly confirmed by international rankings (World Economic

Forum, 2018), which put the country among the best-performing economies worldwide. This strong record is primarily owed to Switzerland's openness to the world. The same assertion can be made about the Swiss higher education system and particularly about the two Federal Institutes of Technology, which not only excel in the scientific rankings, but belong to the most international universities around the globe.

Close to 70% of all faculty members and more than 70% of all doctoral students at ETH are non-Swiss. The research network of ETH numbers more than 9,000 international contacts, of which more than 50% are within Europe. This is to say that Europe remains hugely important for the university, and full and unhindered access to the European Research Area is imperative for ETH and the other Swiss universities. While Europe is preparing for the next seven-year Research Framework Agreement (Horizon Europe, 2021-2027), Switzerland's position is still uncertain. Its status will depend on the outcome of the political discussions on a Framework Agreement between Switzerland and the European Union.

The Swiss science community already suffered negative consequences in 2014, when Switzerland was temporarily excluded from Horizon 2020. There is growing concern that Swiss universities could again pay the price for political disagreement between the EU and the Swiss government. To continue this train of thought, Europe may lose some scientific heavyweights should British universities be barred from full access to Horizon Europe because of Brexit, with Switzerland relegated to the rank of a third-party country. This is in the interest of neither Switzerland nor Europe.

Quality through autonomy

A previous section has already alluded to the political framework within which ETH Zurich operates. It is thanks to the political wisdom of the Swiss government and parliament that ETH (together with the whole ETH Domain) has in the past benefited from its autonomous status. This autonomy gives ETH the necessary leeway to determine the direction of the university and the flexibility to seize unforeseen opportunities. Swiss politicians and the supervisory authority, of course, evaluate on a regular basis whether performance targets have been met and the university is prudently managed. But there is no "industry policy" (as there is in other countries) that would prescribe what research fields the university should engage in. ETH's autonomous status is one of its success factors and should not be compromised in any way.

As a publicly financed university, the bulk of ETH Zurich's funding — roughly 70% — comes from the Swiss Confederation. The rest of the budget is composed of third-party resources, mainly competitive research funding.

Compared to peer universities abroad, ETH Zurich has minimal reserves to compensate for a decrease in public funding. Federal funding of research, education and innovation falls into the category of non-committed expenditure, which means that the ETH Domain is more likely to be affected by budget cuts in times of financial austerity.

CONCLUSIONS

Reconcile tradition with the future

Traditional universities have grown into centres of excellence and innovation thanks to liberal and democratic systems. They are built on the legacy of more than 2,000 years of Western civilization and "stand on the shoulders of giants", to paraphrase Bertrand de Chartres (Wikipedia, 2019). Universities can only thrive and contribute to the progress of humanity in a climate of academic freedom and autonomy. If they become the extended arm of a government agency or a powerful corporation, their very core is at peril. What at first sight seems to be a given, is no longer self-evident in times of growing political pressures, scepticism toward science and the concentration of technological power in the hands of a few tech giants. Universities must stand up for their rights and fundamental values. It should not come as a surprise that Jonathan R. Cole, long-time provost of Columbia University, lists academic freedom, along with free enquiry and trust, as the most important core values for any academic institution (Cole, 2016).

Digital transformation is radically changing every aspect of human activity, such as the labour market. In the face of so many unknowns, education becomes a lifelong task. Universities such as ETH Zurich have something to offer for the next generation. Not only do students get a rock-solid education in mathematics and natural sciences, but they are also exposed to critical, creative and ethical thinking, which prepares them for the future.

The global higher education market is a lucrative target for players pushing new business models. The traditional universities would be well advised to take the new "kids on the block" seriously and look more closely at what they can learn from them. On the other hand, traditional universities also face expectations from politicians, the taxpayer and the media, who increasingly demand transparency and accountability. Universities must prove their usefulness in helping solve the huge global challenges expressed in the Sustainable Development Goals of the UN, and pro-actively pursue an open dialogue with society.

The relationship with industry is a delicate one. Collaboration with companies both at national and international level is undoubtedly crucial for speeding up the innovation process, and every party benefits from real partnerships. But effective collaborations need clear rules and mutual under-standing of each partner's particular role.

If universities such as ETH Zurich manage to strike the right balance between tradition and change and are willing to update their "operating systems", they will remain competitive in the global race for talents and will continue to play a crucial role in the progress of humanity.

REFERENCES

Cole, J. R. (2016). Toward a more perfect university, Public Affairs, New York.

DeMillo, Richard A. (2015). "Revolution in Higher Education", MIT Press, Cambridge, Massachusetts.

Die Volkswirtschaft. (2019). "Das asiatische Zeitalter", ("The Asian Era",) no. 1-2/2019, p.10.

ETH Zurich. (2019). Press release "We should treat each other with respect", available online at: https://www.ethz.ch/en/news-and-events/eth-news/news/2019/03/ measures-leadership.html. [Accessed: 3 May 2019]

Fischer, S.-C. & Wenger, A. (2019). "A Neutral Hub for AI Research, CSS Policy Perspectives", available online at: https://doi.org/10.3929/ethz-b-000332541

Mahbubani, K. (2018). *Has the West lost it? A Provocation*, Penguin, London.

MIT Technology Review (2018). "Google's Long, Strange Life-Span Trip", available online at: https://www.technologyreview.com/s/603087/googles-long-strange-life-span-trip/ [Accessed: 15 December 2016].

OEDC (2019). "Measuring and assessing talent attractiveness in OECD countries". Available online at: https://dx.doi.org/10.1787/b4e677ca-en

Schweizerisches Bundesarchiv (2019). Swiss Federal Archives, German only), Available online at: https://www.amtsdruckschriften.bar.admin.ch/viewOrig-Doc.do?id=10001317 [Accessed: 16 August 2019].

The Lancet. (2019). "China's research renaissance", available online at: https://doi. org/10.1016/S0140-6736(19)30797-4

Times Higher Education. (2019). "World University Ranking", https://www. timeshighereducation.com/student/best-universities/most-international-univer-sities-world

Washington Post. (2018). "China's Application of AI should be a Sputnik Moment for the U.S. But will it be?" 6 November 2018.

Webb, A. (2019). *The Big Nine*, Public Affairs, New York.

Wikipedia. (2019). Adage attributed to Bernard de Chartres: "Dwarves sitting on the shoulders of giants" (Nanos gigantum humeris insidentes), available online at: https://en.wikipedia.org/wiki/Bernard_of_Chartres [Accessed: 16 August 2019]

World Economic Forum. (2017). "Realizing Human Potential in the Fourth Industrial Revolution", Available online at: http://www3.weforum.org/docs/ WEF_EGW_Whitepaper.pdf

World Economic Forum. (2018). "The Global Competitiveness Report 2018", available online at: http://reports.weforum.org/global-competitiveness-report-2018/ [Accessed: 16 August 2019]

CHAPTER 17

Maintaining excellence in unstable times

Leszek Borysiewicz

The topic of this presentation is immediately current, even as I write this short summary. We sit at the beginning of May 2019 in the UK without any resolution to the long-running national debate that surrounds Brexit. This debate has polarized opinion in the UK and is heated as it pertains to the core of the nature of the country the UK is to be. As such it has instituted a paralysis in many of the normal activities of government as it has become such a central overarching issue. Therefore, this single issue has resulted in instability that has dominated the internal debate and, in the UK, exemplifies the uncertain external environment that the Higher Education sector faces yet is limited in the way it can influence outcome. However, further examination of the issues faced by Higher Education Institutions immediately identifies further uncertainties that amount to the development of a perfect storm. The dominance of the debate around Brexit results in other issues failing to be addressed or debated because they are crowded out of media/public consciousness. This further restricts the very limited capacity influence events by HEIs. Paradoxically, such uncertainties are faced by Universities in other countries, but seem more acute in the UK because of a background that threatens a more isolationist environment. For the purposes of this discussion I will concentrate on how these instabilities influence the environment in which this vital sector for the UK national interest has to operate, but I will also refer to the situation in many EU countries, particularly those in Central Europe where I have encountered them. However, as we may see through the lens of Cambridge University, such turmoil is nothing new to Institutions with a long history!

THE ROLE OF A UNIVERSITY

Cambridge is a long-established University, formed as a result of scholars leaving Oxford in 1209. The continuous genealogy of universities, although often attributed to ancient times of Greece and Rome, realistically begins in medieval Europe, and with the Church. In Bologna, Paris, Cambridge and Oxford, the duty of the medieval university was to prepare leaders for the Church and for public life. However, the advent of scientific studies began to bring out a new function which was not just education of undergraduates destined for administration, law and the religious life, but engendered the spirit of discovery and ultimately translating those new ideas and discoveries into benefit of society. This begs the question of what constitutes "society". In earlier days, society was restricted to privileged groups — the state and church in particular. However, this quickly became the community in the local vicinity of the University, gradually expanding to the nation. And some today remain locked into this concept, yet most academics in Universities worldwide view today's world as a single society and therefore beneficiary of discovery and new ideas. This inherent internationalism has placed the Universities at odds with a prevailing position of "leaving the EU"; it is interesting to note that not a single HEI (of approximately 160 such institutions) supported the "leave" campaign in the recent referendum. A unanimity that, I suspect, has never before been achieved!

The 19th century was busy for Higher Education in the UK and much of Europe, but it led to a number of thinkers opining and developing the underlying philosophical framework for the purpose of Universities. Appreciating these concepts is important as it emphasizes the differences between UK and Continental European Universities and has led to many misunderstandings in the debates on Higher education in the EU. Wilhelm von Humboldt in Germany and Cardinal John Henry Newman in England and Ireland set out competing and overlapping Ideas of what universities should be for, building on, rather than demolishing, the medieval idea. By a quirk of fate and global politics, Britain rejected the development of Universities as institutions linked by religion (this was largely rejected in the 19th Century) and the consequences of the European "Free University" (i.e. secular University) movements, as well as the separation of teaching and research into separate Institutes. This also transplanted to University systems throughout the then British Empire as well as the US, which explains fundamental differences between EU systems of Higher Education. In the 20th century, the unification of teaching and research in universities, following Humboldt, became the common paradigm in the UK and US rather than specific research Institutes. This continues today in the UK. The most recently established Research Institutes are all linked with Universities. For example, the Crick Institute in central London encompasses the former independent London CRUK Cancer Research Institute, the MRC National Institute for Medical Research at Mill Hill, but unites it with University College, Imperial and King's College London.

Furthermore, universities began operating on a global stage in keeping with their acceptance of a paradigm of global society. My point is simply that in every historical and geographical incarnation of a university, "making a difference in the world" has been a recognizable aim although prevailing national influences have coloured how this is projected externally. Ultimately HEIs do not operate in a societal vacuum!

But there are core principles that are espoused by Universities, wherever they are found. Central to these is the principle of "Academic Freedom" — the ability of individual academics and students to freedom of thought and investigation to enable them to develop new concepts and discoveries. However, academics do not withdraw into universities to think deep thoughts — they deepen those thoughts by constant engagement with others, hence the deep held conviction of the freedom to collaborate freely. Universities, though sprung from monastic roots, are not monasteries — they are functionally the opposite.

This concept is well enshrined in the mission statement of Cambridge University, only one sentence long:

"The mission of Cambridge University is to serve society by teaching, research and learning at the highest international level."

Therefore, there is an implicit contract between society and Universities: society endows Universities with privileges, such as "Academic Freedom" and "Institutional Autonomy" because there is **trust** that their use of these freedoms will generate societal benefit.

However, society, especially national society, has placed far greater demands on the purpose of a University than originally intended and this is given greater prominence because Universities are often supported by public funds. Universities are tasked by society through governments and countries that support them with multiple objectives: to educate the population of the host nation to an advanced level; to promote social mobility in that nation by providing a level playing field for access to that education, regardless of social background; to make new discoveries through research and thereby push back the boundaries of human knowledge; to act as custodians of knowledge and of culture; and of course to generate income for the country, by attracting overseas students and by making useful and patentable inventions which in turn result in wealth and job creation.

In some ways the miracle is that most Universities deliver on all these goals, sometimes with more emphasis at an individual institution on one or more of these, but as a sector it delivers on most maintaining the delicate concept of public trust.

How does Cambridge deliver on these goals?

1. **Education.** This is first and foremost the function of a University — to build up the next generation who in turn will build the future. New ideas stem from "standing on the shoulders of giants", a phrase used by many academics to describe how they attained their achievements. But the education provided is distinctive and different at each University albeit with a common goal. I believe this variability is a fundamental strength of higher education rather than a weakness. It allows for choice by the student of the course of study that suits their own goals best. Yet this approach is expensive. Cambridge has a unique (alongside Oxford) method which is based around the University and its constituent Colleges. Undergraduates apply to the University by choosing a course of study e.g. history, at a specific College. On entry the student receives instruction at the University in terms of formal teaching (i.e. lectures or laboratory studies) and is examined receiving their degree from the University while the College provides small group teaching (often 1:1) to supplement and enhance the formal education. This is a hugely intensive and thus expensive undertaking — the average cost of a year's instruction to the University and College is £19,000, yet the government will only provide the student (UK and EU) with a loan of £9,000, which is also the maximum the University can charge. The difference of approximate £80 million each year has to be made up from other sources — mostly our endowment. Financial management can just manage this, but it causes conflict if government would seek to interfere with the admission process or course content/duration — after all it doesn't pay for it in the UK system!

2. **Postgraduate/postdoctoral studies.** Nearly all Universities recognize that higher education will not end at undergraduate level but requires further study — taught Masters and research-led PhDs. There is growing demand for more of these qualifications — while Cambridge has 11,000 undergraduate students, there are an additional 6,000 postgraduate (4,000 PhDs and 2,000 taught Masters) and nearly 4,000 independently funded postdoctoral researchers. Responsibility for these communities is vital for their development as experts in their fields but also because of the national need for their skills.

3. **Social mobility.** The demand for places at a University such as Cambridge results in intense competition at undergraduate and postgraduate level; only 20% of applicants are successful in their application at Undergraduate level. Many of the unsuccessful students

will achieve the academic standards through examination yet will not have entry. So how to create opportunity for those from disadvantaged backgrounds is a key question — recently the University announced a call to raise special funding to support such individuals.

4. **Discoveries and New Knowledge.** Most of the world-leading institutions are recognized as such not through their excellence in teaching (which most academics still see as their primary function) but through research output. Therefore, great care must be exercised in interpreting so-called league tables as these are dominated by what is easily measurable rather than the full mission of a university. Research output is easier to evaluate — in fact there is not a single credible internationally validated measure of teaching excellence! It is also the major source of funding to such Universities and largely what attracts the best international staff. Of the total annual turnover of the University of £1.5 billion per annum, nearly £450 million is through competitive grant awards by government and charitable foundations. This is also a source of great pride to the University — in 2018 we celebrated our 97th Nobel prize to Greg Winter for phage display and humanisation of monoclonal antibodies for human utilisation. However, it places emphasis on research as the major criterion when academic staff are appointed, but all these staff from the youngest Lecturer to the Nobel prize winner are expected to teach and supervise! Yet the pursuance of "new knowledge" be it in philosophy through Wittgenstein or new drugs through Winter, not only fulfils the Humboldtian vision of a shared responsibility of student and academic to seek new knowledge but delivers the unwritten contract of benefit to society — not just economic but also social.

5. **Generation of Intellectual Property and economic wealth.** Cambridge University, through its creation of and engagement with the Cambridge Phenomenon, has developed Europe's largest industrial cluster. There are currently over 4,000 companies within a 20-mile (32km) radius that build on the know-how of the University; 15 of these now are valued at over £1 billion and ~4 at > £10 billion. Cambridge is small with a population of ~120,000 and a surrounding population of 600,000, yet 17% of all high-tech startups in the UK happen here, and between them they have created 60,000 jobs. They attract multinational research companies such as Microsoft and AstraZeneca and contribute £13 billion per annum to the UK economy. This is considered a huge national success, but it does not happen overnight. This is based on 50 years of development, investment and belief in the importance of fundamental studies that eventually translate and a *laissez-faire* approach that does

not pre-define disciplines or domains but allows the opportunity to all. There is a widespread view that external imposition of structure would destroy rather than enhance the Phenomenon. Maybe a success of chaos over order?

6. **Repositories of Infrastructure, Knowledge and Culture.** To maintain this approach to education, learning and research requires a considerable investment in maintaining an expensive infrastructure. This consists of libraries, some dating from the middle ages as repositories of knowledge, as well as University museums (11 in total, the largest being the Fitzwilliam Museum) which all function to support the three principles of the University mission. For some of the Colleges, this also includes UNESCO treasures such as Kings College Chapel. But the biggest expenditure is provision of laboratories, equipment and accommodation within a short distance from the core buildings/ laboratories of the University — the largest such development was to build a new site which adds 15% to the total size of the City at a cost to the University of £1 billion. This highlights the need to invest at scale and risk — possibly the true price of institutional autonomy. Universities have to be sustainable, make appropriate investment decisions recognizing that under the current structures within the UK there is no "safety net". So, autonomy also comes at a price.

If Universities are the mainstay of the UK research effort and have the right to autonomy, they have to be able also to manage risk and uncertainty as well as delivering the academic agenda. This leaves them exposed to uncertainties and at present these abound in the UK and elsewhere.

UNCERTAINTIES

There are inherent global economic uncertainties that Universities face with respect to finances, investments, fundraising, infrastructure, income, but most higher education institutions should be well versed in managing these. This is a global issue for HEIs either directly or as a consequence of available government investment in HE, especially in the face of economic downturn and falling tax revenues. In every country HE has to compete with all the other demands on funds, especially as regards the delicate balance of funding primary and secondary education. However, the clouds of external uncertainties are gathering on the horizon and the scope for HE to mitigate their potential impact is increasingly limited, in a global climate that espouses a dislike of "experts". Turning the uncertainties from challenges/ threats will lie at the heart of ensuring a thriving HE sector for the future. However, the background in the UK is complex.

Firstly, there is a complexity to University funding in the UK that is a consequence of government policy largely derived from the time of the coalition government after 2008. After that economic downturn, it was essential to consider how the costs of a University education were to be met. The previous goal of the outgoing Blair administration was that 40% of the population should access HE. Once established as a benchmark, this is impossible politically to reverse, as exemplified in many countries where universal entry is enshrined in constitutional rights e.g. France, Slovenia etc. The coalition government opted for a "market" solution, which recognized in particular the individual benefit gained by a student from attending University. (Most of us believed that this underplayed the overall benefit to society of a well-educated population!) The solution was to raise student fees from £3,000/year (introduced in 2003) to now £9,250/year by creating a Student Loan Company to which students could apply for a loan repayable once their income was above a threshold of £18,000/year (rising to £25,000 in 2018/19), through the taxation system. This ensured that Universities obtained income but allowed the government to largely stop paying directly through a T grant. There was a major debate as the minor party in the coalition was elected largely on its opposition to fee increases. Unfortunately, this scheme is increasingly uneconomic and growing politically unacceptable:

1. The repayment alongside a higher student drop-out has raised the interest on the loan to students to commercially unsustainable levels to off-set losses.

2. The Student Loan Company is currently in deficit to £12 bn rising to £17 bn in 5 years with a projected 45% failing to repay the debt (2018 — Institute for Fiscal Studies).

3. The original concept that a "market" was going to be created and institutions would compete on price has failed — virtually all Institutions charged the maximum fee. It was negated further by concessions to establish the system on a maximum cap as well as preventing early repayment because of social equity.

4. The removal of student number control for HE entry has not increased competition on price, but increased the deficit of the SLC.

5. Student fees are politically toxic. The minor party in the coalition was almost annihilated in the 2015 Election and the current government attributed the success of the opposition in the 2017 election to be due in part to a promise to cancel student debt and abandon student fees — something that they have now withdrawn as the costs of adding the SLC debt to the Treasury would be catastrophic.

But this has resulted in several fundamental changes:

1. The perception that all Universities in the UK are "private".
2. The government wishing to control HE but at the same time not being willing (or able) to pay the real costs of HE, establishing a conflict.
3. The creation of a market and commoditization of HE — the student as a consumer. This is seen by government as being akin to owner-ship of the system of HE by the "consumer" who with their "share-holder" pressure will drive price down while increasing quality. This challenges a key Humboldtian principle that student and teacher work together to further the acquisition of new knowledge. This has caused debate about the purpose of HE among academics, with a perception that we are creating a situation where, at its extreme. the only outcome of HE is salary and not broader contribution to society e.g. FT League Tables for MBA.

Secondly, this first uncertainty is now compounded by continuing reviews and potential further changes. As I write this paper, we are awaiting the final report of the Augar Review on Post 18 Education, possibly as soon as next week. If, as widely trailed, it will recommend a reduction in the cap of anywhere between £6,500 to £7,500, this will significantly impact on most Universities' income. Furthermore, this is in a climate where there is no certainty that Treasury will be in a position to re-institute an increased T budget. It remains unknown if student number or quality control will be introduced as an eligibility criterion and an even bigger question remains as to the parlous state of funding for Further Education Colleges.

Thirdly the government instituted a review and ultimately presented and passed the Higher Education and Research Act in 2016. This has established far-reaching reforms, which are fundamental to the climate in which HE oper-ates in the UK. While all Institutions have acted as if there is formal Institutional Autonomy, this is now fully recognized as is the Dual Support System which ensures that Universities receive funding to support research, they undertake that is externally funded. In addition, a longstanding principle in the UK — the Haldane Principle — has been formally recognized (that research funding is [relatively] independent of political interference). However, among espousal of these fundamental principles, there have been other major changes:

1. The abolition of the Higher Education Funding Council (an "arm's length" body that distributed government resource to Universities but also sought to maintain equanimity in the sector e.g. helping HEIs in financial difficulties).
2. The replacement of HEFCE with a "regulator" — the Office for Students. This has changed the whole basis of interaction with

Universities and brought numerous agencies such as the Office for Fair Access, complaint management etc, under a single entity, but one with a "consumer" focus rather than a body that worked in collaboration with the sector. How this will play out in the longer term is very uncertain, and concerns have been expressed about the real independence of this regulator.

3. The establishment of a Register of Universities with as yet non-defined quality measures. This ushered in a Teaching Excellence Framework (akin to the Research Excellence Framework) but without the financial benefit of the latter for excellent performance! Again it is unclear what further measures will be instituted.

4. Opening the "market" to "new providers" by using the Register. This is largely seen as an effort to increase competition in the sector and reduce costs to students. Neither is evident as yet.

5. The longstanding duality of Teaching and Research has been broken with a split of the two functions between government departments; T residing in the Department for Education and R with Department for Business, Energy and Industrial Strategy (sic).

6. Government research funding has been reformed along the lines suggested by the Nurse Review. The seven Research Councils alongside Innovate UK (a body that supports and develops SMEs often associated with Universities) and Research England (which provides the quality-based research support for English Universities through the Research Excellence Framework) are all brought under UK Research and Innovation, an independent body that will advise on relative funding allocations between these nine entities. The government has made two significant promises: firstly, increasing the R&D budget by an extra £2bn/year to £8.6 bn and secondly, to set a target that the UK would invest 2.4% of GDP in R&D. While this is significant, much debate has ensued as to how far the new resource is being used to support a central plank in government policy — the Industry Strategy — rather than ensuring a balanced basic vs applied research portfolio.

National uncertainty, and these very significant changes, creates a difficult environment for Universities to chart a course of fiscal and principled probity. The central issue of societal trust is significantly threatened as public opinion for a variety of reasons perceives Universities as privileged, rich and a root cause of endangering social mobility.

To merely address the financial, Universities would need to consider: where cuts would need to be made; investing at risk in increasing student numbers; or expanding courses, without increasing delivery costs. However, any of these responses is likely to result in reduced student satisfaction.

Alternatively for some, new models of approaching higher education through on-line or two-year courses (46 week study/year) or even complete independence will be considered. However, as a positive it may also herald rapid change with a greater espousal of new technologies to overcome some of these difficulties. The options to increase size need to take heed from the experience of countries where there is universal access based on performance in secondary school exit examinations. Class size is so large and loss of a percentage of students after 12 months at University create staff dissatisfaction that is evident in France and Slovenia where the appropriateness of this model is under debate.

Pensions. Institutional autonomy, as practised in the UK, requires the University to behave as a "private" employer. This requires the provision of a pension scheme for all employees. Academic staff largely fall under a mutual and exclusive scheme (Universities Superannuation Scheme — USS), which is in significant deficit. Projected is a large increase in employers' contributions which will add many millions to the salary bill. Where are the cuts to be made to make up this shortfall? How will this be accommodated — potentially job losses or failure to take on new staff may happen or again increasing class size in teaching orientated Universities.

BREXIT. As I compile this discussion paper, the announcement of the resignation of the Prime Minister has been made. For many outside the UK and EU, the deep division this debate has created in the body politic and the country at large is difficult to conceptualize. Whichever side of the debate individuals stand on, there are such fundamental forces at play that the divisions in society may take a generation to heal. Universities were (uniquely) unanimously opposed to Brexit and therefore find themselves on one side of the debate. The UK remains in limbo.

Debate has focussed on the question of, if we leave, then under what terms will this happen. Academics largely support a position that they largely oppose leaving, but if this were to happen then the closest possible association with the EU as regards R&D funding should be sought. However, the nature of associate country status causes considerable debate, with the alternative "no deal" or breakaway scenario vacillating as a likely outcome. The concern is that the UK's very success in R&D funding will not be fully recognized and resources will not be made available to the sector on the same scale. The factors at play here are both competition for an ever-dwindling resource that remains of the monies that would be repatriated from the EU (between large sectors such as fisheries and agriculture) and a predicted economic downturn that will require emergency support in other areas with R&D missing out. Perhaps even more worrying is the lack of infrastructure investment by the commercial and public sector since the referendum, eroding the UK's competitive position while these debates play out!

Quite separately from the political dimension the consequences will be far-reaching especially with respect to R&D. The UK receives the largest share of ERC and a very large share of all EU funding in R&D. UK HE institutions have enjoyed consequent collaboration with many European centres. Most telling is the observation that other EU countries now form the largest group of collaborators (rather than the US) by publication. The possible financial loss will probably be partly recoverable, but the academic loss to the UK would be huge. To date the politicians are committed to the "closest possible" links with the EU, supported by so many academics in the EU, but with the current turmoil, who knows?

Immigration and competitive recruitment. This cannot be disassociated from BREXIT. However, limiting immigration is a particular problem especially when so many of our best investigators are international. Any severe restriction would impact negatively on the ability of UK institutions to compete globally, but even the perception of hindrance to movement will have a negative impact.

Being independent and competitive between themselves, UK Universities compete globally to attract the best academics. However, this leads to considerable and spiralling salary inflation as the competition at the highest level is with well-endowed US Institutions. Will this result in a narrowing base of institutions able to compete? If so, alongside the other financial pressures, it will challenge the viability of some Universities, and mergers and acquisitions may start to occur in the sector. Most observers are concerned that reducing the number of Universities would reduce diversity and opportunities for staff and students.

Trust. As always a major concern in the UK as elsewhere is the issue of societal trust if it were to be undermined by these debates. In surveys of trust, universities and academics have and continue to perform well as opposed to the media and politicians who are almost universally distrusted. But the impact of social media, vilification of expert opinion and populism, all of which are counterintuitive to the HE cultures we strive to engender, may take their toll. The sense of Universities as rich, self-indulgent and privileged is real and must be countered so that we do not lose this vital compact. Issues such a vice chancellor's pay, value for money, openness, and relevance require us to engage with this debate and not assume that it is a given.

CONCLUSIONS

Many of the issues facing the UK have their counterparts in the EU and the rest of the world. The UK is in some turbulence at the moment, but elements of these trends are evident in other countries. Financial pressures are universal, the public debate of Universities' role in and for society widely tested. The impact of commercial interests ranges from seeing these as a salvation to a threat to academic freedom. I suspect that the debate will play out differently in different countries and it is impossible to predict where in the spectrum of solutions the UK will find itself.

To further complicate matters, there are many other risks in the longer term that could be added to this list and the paper could become very negative. However, wherever there are challenges there are also considerable opportunities and the determination of the academic and University sector will be to stay true to its principles, seek the widest possible international engagement and look to develop new ideas and discoveries for the benefit of society. The current situation may be different but the message from history is optimistic. Universities are among the most enduring of social structures. In fact, alongside religious institutions they are well ahead in terms of longevity compared with any commercial concerns or even systems of government. Universities have survived and thrive through worse than the current uncertainties, — in the case of Cambridge, the Reformation, Counter-Reformation, Counter-Counter-Reformation, civil war, global conflicts — and still remain world-leading institutions that are valued for the diversity of their functions and continue to serve society. I firmly believe that this will be the case in the future.

ADDENDUM (ADDED 18 SEPTEMBER 2019)

Since the manuscript was prepared little of substance has changed for the UK. The political turmoil surrounding Brexit has intensified with political defeats for the new Prime Minister and a decision by Parliament that a "no-deal" Brexit will not be supported. Calls for a general election, expulsions of objectors from the ruling party who would not support "no-deal", failure by government to force a general election and even a case in the Supreme Court assessing the legality of moves by the government in suspending Parliament have intensified the debate rather than resolved it. The complete focus on Brexit has largely resulted in little movement on the other key issues raised in the paper, and I am sure that this will develop only later in the year.

Throughout, the EU has been consistent in asking what the UK administration wants in terms of a settlement for Brexit but to date no specific proposals have been forthcoming. Therefore, the sense of limbo continues, which is unlikely to result in progress on the issues that the HE-sector faces in the UK. Ultimately, these will have to be resolved but the view remains that none of this will be addressed until Brexit is resolved.

FURTHER READING

Collini, S. (2012). *What are Universities for?* Penguin, London pp. 240.

The Royal Society. *UK research and the European Union: the role of the EU in funding UK research.* Online:
https://royalsociety.org/-/media/policy/projects/eu-uk-funding/uk-membership-of-eu.pdf

The Royal Society. *The role of the EU in international research collaboration and researcher mobility.* Online:
https://royalsociety.org/-/media/policy/projects/eu-uk-funding/phase-2/EU-role-in-international-research-collaboration-and-researcher-mobility.pdf

The Royal Society. *The role of EU regulation and policy in governing UK research.* Online:
https://royalsociety.org/topics-policy/projects/uk-research-and-european-union/role-of-eu-regulation-and-policy-in-governing-uk-research/

Sweeney, D. (2019). "Building partnerships and trust". *J. Foundation Sci and Tech* 22; 33.

Rothwell, N. (2019). "A force for public good". *J. Foundation Sci and Tech* 22; 35.

Willetts, D. (2019) "Institutions with a host of roles". *J. Foundation Sci and Tech* 22; 36.

CHAPTER 18

Technology and Humanity for Industry 4.0 and Learning 4.0

Subra Suresh

S cientific discoveries and engineering innovation are accelerating the unprecedented convergence of the physical, digital and biological worlds to produce technological advances that are poised to disrupt and transform the daily lives of ordinary citizens at an ever-increasing pace [1, 2]. This ongoing transformation has been broadly and commonly referred to as the Fourth Industrial Revolution or Industry 4.0 [3].

The disruptions engendered by this revolution have been catalysed by developments in many research and applied fields. They include, but are not limited to:

1. computing hardware and software;
2. massive and deep data analytics;
3. blockchain;
4. mobile communication technologies, augmented in the future by 5G;
5. autonomy and intelligence of machines and robots;
6. advanced additive manufacturing;
7. personalized medicine;
8. augmented and virtual reality;
9. industrial internet of things;
10. genomics, gene-editing and computer chips augmented with genome-analysing features;
11. nanotechnology; and
12. metrology enabling improvements in precision and resolution with which time, location, as well as physical and chemical properties and characteristics of matter and objects, can be measured.

TECHNOLOGY AND INDUSTRY 4.0

The First Industrial Revolution, originating in Scotland in the 18th century, was propelled by the mechanization of labour by recourse to steam and water as energy sources which replaced human and animal labour. The Second Industrial Revolution, also commonly known as the Technological Revolution, which evolved from the late the 19th Century until World War I, was marked by advances in electrification, factory assembly lines, machining, rail transportation, metal processing, manufacturing and telegraphic communication. Industry 3.0 was catalysed in the 20th century by advances in microprocessors and computing, automation, robotics, programmable logic controllers and the evolution of global supply chains.

By comparison to the previous three industrial revolutions, Industry 4.0 is marked by a number of unique characteristics:

Figure 1– The time required for different technologies to mature and to be adopted by the first 50 million users. Data courtesy of *The Wall Street Journal* and Valuecapitalist.com.

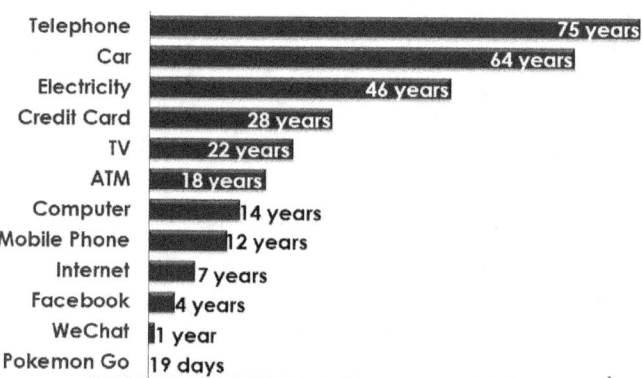

Source: Time to reach the first 50 million users

1. The pace of technological change and disruption has never been faster and the ensuing implications for individuals and societies have never been more pronounced. Figure 1 is an illustration of the accelerating pace of innovation and its widespread adoption of technologies around the globe. Their transformative effects impact the lives and livelihoods of billions of global citizens.

2. Aspects of normal and routine human endeavour will be increasingly influenced by decisions made by machines with real-time access to massive amounts of aggregated data. Such decisions are expected to have a profound impact on the daily lives of humans in areas as

diverse as transportation, medical diagnosis and treatment, managing personal well-being, manufacturing, logistics and supply chain, assisted living, cradle-to-grave education and learning, delivery of healthcare, and care of the elderly, those with special needs, and children. Industry 4.0 has accelerated bi-directional communication between the individual citizen of the world and the leading edge of disruptive transformation by recourse to mobile technologies. This trend is poised to see a major leap in coming years as 5G communication technologies will be rolled out in the not-too-distant future.

3. Personalized machine learning algorithms, incorporating either unsupervised or partially supervised learning, are also used to target individuals, organizations and communities to automatically flood them with information/misinformation at speeds much faster than properly vetted, reviewed and authenticated real news can travel. These technology-enabled communication channels employing websites and social media often intentionally obfuscate the unsuspecting target by trumping truthful information with sophisticated and seemingly authentic fake news or information predicated on biased data and statistics. They can also instigate political, financial, commercial and even physical harm to citizens, communities and countries.

4. Industry 4.0, unlike at any previous juncture in human history, raises fundamental questions about the potential for humanity to be altered by technology. It also raises concerns about the degree of long-term irreversibility associated with the influence of technology in such areas of societal importance as climate change, sustainability of the planet and of the quality of life of its inhabitants, equality of income and opportunities, fairness, ethics, risk, liability, regulations, responsibility and governance.

INDUSTRY 4.0 AND HUMANITY 4.0

The foregoing unique characteristics of Industry 4.0 raise many fundamental issues and challenges for humanity. Here we pose six major questions. How societies address these issues individually and collectively will determine whether technological advances influencing the fourth industrial revolution will ultimately turn out to be net positive or net negative for humanity.

1. Every previous industrial revolution resulted in massive job losses, but it ultimately (and, in most cases, over a span of several decades) led to the creation of more jobs than the number of jobs eliminated. In Industry 4.0, with an unprecedented pace of anticipated rapid societal change, a long time delay between the elimination of

current jobs through the wider adoption of "intelligent" machines and the creation of new jobs is expected to further accentuate the growing disparities in income and in quality of life among citizens of many countries. This could also lead to further polarization of countries and societies toward extremes.

2. A high school or university graduate today is expected to continually learn to adapt to the transformative changes created by technology. Today's graduate is also expected to change jobs and even professions many times over the course of a long career. In order to succeed in the increasingly competitive global marketplace driven by greater efficiency, what is the "minimum body of knowledge" a university graduate is supposed to acquire during formal education so as to be prepared to acquire new skills over a lifetime of rapid changes in workforce needs? What are the roles and responsibilities of educators, employers and governments in providing these basic skills not only during the early years of formal learning and employment, but also for continual "re-skilling" and "upskilling" for "lifelong learning" throughout one's career and life? What does it mean to be "an educated person" in the 21st century?

Figure 2– Six of the key issues and questions surrounding Industry 4.0 as human behaviour interfaces with technological advances and with disruptions arising from such advances.

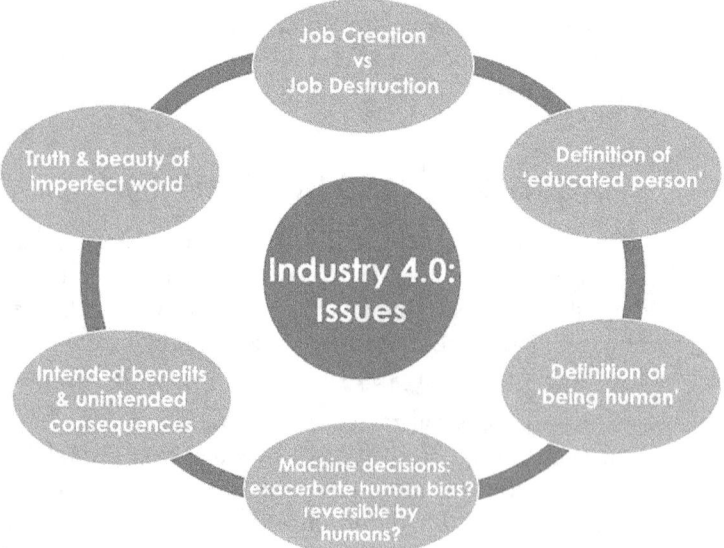

3. As noted earlier, technologies enabled by advances in such fields as computing, artificial intelligence, machine learning, and real-time and deep data analytics are poised to play an important role in determining, influencing and controlling a vast spectrum of human endeavours and activities. At the same time, distinct differences arising from cultural, social, national and family circumstances, along with individual life experiences, uniquely shape the evolution of non-duplicative characteristics of each human being. Manifestations of individual uniqueness among billions of people lead to such distinctly human characteristics influenced by personal values such as dignity, ethics, empathy, compassion, sympathy, pride and honour. With decisions made through the agglomeration of massive amounts of data, will machine decisions begin to influence human activities in a manner that distorts innate individual characteristics and values and the ensuing behaviour patterns? Whose algorithms and perspectives will determine values that are important to an individual, on what basis, and relying on what kinds of data? Who will authenticate and vouch for the veracity of such data? In other words, in the era of Industry 4.0 in which human actions and activities are expected to be increasingly influenced by machine intelligence and decisions, what will it mean "to be human"?

4. There is existing evidence that biased data input or algorithms for neural networks and machine learning, involving such technologies as face recognition, can sometimes lead to bad, unacceptable, and even "evil" decisions. Will machines help mitigate or exacerbate innate human biases, whether conscious or unconscious, through bad, erroneous, unreliable or insufficient data? Under what conditions can machine decisions become irreversible and permanent making human intervention impossible, irrelevant or immaterial?

5. Many technological advances ultimately lead to the betterment of human condition. Most of them also create unintended consequences that have deleterious effects on humans and society. In 2000, at the dawn of the new century and the new millennium, the National Academy of Engineering (NAE) of the United States released a list of 20 greatest engineering achievements of the 20th century (see: http://www.greatachievements.org/). This list includes impressive accomplishments such as: electrification, automobile, airplane, computers, internet and nuclear technologies. Several years later, NAE also released a report on the 14 grand challenges of the 21st century (see: http://www.engineeringchallenges.org/). This latter report includes such global challenges as: restoring and improving urban infrastructure, securing cyberspace, providing access to clean water,

preventing nuclear terror, and developing carbon sequestration methods. When we examine the two lists side by side, we cannot help but wonder whether some of the greatest engineering achievements of the 20th century played a pivotal role in creating some of the toughest grand challenges for the 21st century. The greatest engineering achievements of the last century led to enormous benefits to humankind and elevated quality of life around the globe. At the same time, in the course of solving some of the hardest technological problems to produce innovative products that led to many tangible benefits to society, we created some of the most difficult challenges and unintended consequences for succeeding generations. Then, how likely is it that our even greater technological accomplishments of the 21st century driving Industry 4.0 will not lead to even grander challenges for the 22nd century? What was missing in our collective thinking in the last century that needs to be addressed now so that we do not repeat our past mistakes in this century?

6. Technology has advanced to a level of sophistication whereby Global Positioning System (GPS) can pinpoint a location with real-time kinematic positioning to centimetre-level resolution (https://en.wikipedia.org/wiki/Real-time_kinematic). Atomic clocks routinely monitor time to a level of temporal accuracy whose error rate is better than a billionth of a second per day (https://en.wikipedia.org/wiki/Atomic_clock). Transmission electron microscopes now routinely provide clear images of individual atoms in materials with spatial resolution on the order of 0.1 nanometre. Personalized and individualized genetic testing of DNA from a saliva sample and associated data analysis can provide ancestry estimates down to 0.1% of global population and gene pool (https://www.23andme.com/en-int/). Technological advances place increasingly greater emphasis on precision, perfection and prompt action in many human activities where they are deployed and adopted on a massive global scale. This trend has nurtured a relentless and ever-accelerating pace of work that encroaches on personal time and space, driving ever-greater precision, perfection and immediacy of action. However, truth and beauty associated with imperfection and imprecision, deliberate allocation of sufficient time for relaxation, meandering, exploration and reflection, and the notion that failure and imperfection are a necessary part of the learning process, are also known to be essential ingredients for nurturing artistic creativity and scientific discovery. As technology forces individuals and professions toward greater degrees of precision and perfection in Industry 4.0, what are the consequences for human behaviour in an intrinsically imprecise and imperfect world?

The foregoing complex questions and issues require collective thinking and action across professional, disciplinary, geographical, intellectual and national boundaries. First and foremost, these issues are not just engineering or technology-based issues. They are also strongly predicated on human behaviour. It is perhaps prudent to consider first how human psychology, values, aspirations and limitations will intersect with emerging technologies and their anticipated massive disruptions arising from Industry 4.0. They must include concerns about climate change, sustainability of natural and renewable resources, concentration of as much as 70% of the world population in urban areas and mega-cities, growing inequality in income, wealth and opportunities within and among populations, and the increasing role of machines and their real-time decisions affecting a vast array of routine human activities.

INDUSTRY 4.0 AND LEARNING 4.0: SOME CONSIDERATIONS FOR UNIVERSITIES

Now we consider a few ideas for tertiary educators and universities that could help address some of the issues raised in this paper. Although most of these perspectives are not new, they connect to the challenges discussed above.

A. A critical assessment of the "basic skills" taught in university curricula is needed to prepare students to adapt to a lifetime of technological and societal transformations catalysed by Industry 4.0. Specifically, what special skills does an undergraduate student need to acquire at a university in a time frame that is no longer than four years? What should be the required minimum set of courses and subjects across disciplinary boundaries to prepare the student for a lifetime of re-learning, up-skilling, productive citizenship and a purposeful life? How do different fields as diverse as the arts, humanities, social sciences, business and economics, medicine, natural sciences and engineering assess such needs for basic skills? What is the minimum body of knowledge that a university graduate (an educated person) of the 21st century should possess? As a first step in this direction, Nanyang Technological University (NTU) Singapore introduced minimum course requirements in "digital literacy" (which also includes such topics as ethics in the digital age) for all of its more than 23,000 undergraduate students, beginning with the incoming freshman class of 2018.

B. We briefly examined NAE's 20 greatest engineering achievements of the 20th century and the fourteen grand challenges of the 21st century (see Figure 2 and item 5 discussed earlier). Some would argue that perhaps sufficient attention was not devoted to the integration of technology with human behaviour and with humanity in our

collective effort accompanying the rollout of the impressive innovations of the 20th century. Universities could consider formal and
informal ways in which such integration routinely becomes part of
the education process. This will require tighter coupling of natural sciences, computing, engineering and medicine on the one hand
with social sciences, arts and humanities, with topics such as human
psychology, communication, ethics, economics, and governance not
left out of a broader and more complete curriculum for all students.

C. Mobile technologies and digital information increasingly impact
every aspect of human life. Whether a university graduate is an Arts
major or a science major, computing and digital technologies will
increasingly play a pivotal role in the ability of the graduate to function as a productive citizen of society. Given this trend, computing
becomes as much of a "required" subject in a university for an arts or
humanities student, as literature and social sciences should be for a
student of computer science.

D. It is now widely recognized that rapidly expanding academic disciplines such as artificial intelligence (AI), machine learning (ML),
robotics, precision medicine and 3D printing are poised to shape
the course of industry in the coming years and decades. However,
the impact of these disciplines in shaping the lives and livelihoods
of billions of ordinary citizens of the world and in solving some of
global society's most pressing challenges has perhaps been less of a
focus of academic discourse than its economic and industrial implications. Universities have an opportunity, and some would argue an
obligation, to address ways in which the role of these intellectual
disciplines could better the lives of under-privileged citizens of the
developing world. For example, how can AI and ML advances be
used to address the needs of the under-privileged affected by such
issues as pollution, job loss, human trafficking, lack of access to clean
water, paucity of access to banks and fair lending practices, healthcare, information and basic education?

E. Many governments and industries, along with thinktanks and non-profits, have identified ways in which citizens can receive support and
assistance in their efforts to upgrade their work skills. For example, the
government of the Republic of Singapore has rolled out the SkillsFuture
program (see: https://www.skillsfuture.sg/) to provide its citizens opportunities for lifelong learning outside formal educational organizations
and employers. The government has also provided free credits for citizens to incentivize learning. Universities have an opportunity to engage
alumni and citizens, from the region and around the world, to tap into
opportunities to taking courses and obtaining credit. Many universities

have already introduced such mechanisms, from micro-credits to full course credits to online degree programs, with varying levels of success. Nevertheless, there is a critical need to address the issue of aggregating and validating such credits (even for a university's own alumni) that are transportable to employers. This could mirror, with appropriate modifications, pathways for university degrees to be authenticated in many cases by the endorsement of accreditation bodies and governments.

Figure 3– Some strategies for enhancing learning outcomes in Industry 4.0.

Critical assessment of core skills for an "educated person" of the 21st Century

Digital literacy, digital hygiene, social responsibility, commitment to sustainability

Role of social sciences, ethics, and humanities

Forever learning with public-private partnerships

Responsible Innovation

Training to manage information and misinformation

F. Finland has emerged as a country that is most resistant to managing misinformation. The approach adapted there involves education in the classroom about real and fake news and training students and citizens about the importance of authenticity of information for the health of society (see: https://www.weforum.org/agenda/2019/05/how-finland-is-fighting-fake-news-in-the-classroom). Universities can play a vital role in this regard by providing proper education about authenticity of information, critical thinking and reasoning, as well as digital literacy and "digital hygiene".

G. Several universities around the world have created multi-disciplinary activities, centres and institutes to address the intersections of science and technology with humanities, human behaviour, policy and ethics in education, research, advocacy and societal outreach. Perhaps only a subset of such institutions, however, have the scope and infrastructure to engage the full spectrum of stakeholders for successfully translating academic pursuit to societal impact. The stakeholder community should inevitably include government agencies, policy-makers, global industry partners, small and medium enterprises, regulating authorities, and non-profits.

Figure 3 provides a summary of some strategies for enhancing learning outcomes in Industry 4.0.

As a step in this direction, NTU Singapore established in 2018 the NTU Institute of Science and Technology for Humanity (NISTH). This university-wide institute is aimed at bringing together the diverse stakeholder community, in partnership with key government agencies and the many industry partners with a major presence on campus, to address a number of issues and challenges. The three areas of initial focus chosen by NISTH are: responsible innovation; governance and leadership in the era of Industry 4.0; and the new urban Asia.

Figure 4– The three initial areas of focus
of the NTU Institute of Science and Technology for Humanity (NISTH).

CONCLUDING REMARKS

Scientific discoveries and technological advances are creating unprecedented opportunities for individuals, institutions, governments and global society to elevate living standards and quality of life, and to eliminate disparities. At the same time, history has shown repeatedly that intended benefits of technologies are inevitably accompanied by unintended consequences. With the fourth industrial revolution, rapid pace of technology development and mass adoption, along with instant and borderless communication, offer new opportunities and challenges. Educational institutions, working with governments, industries and nonprofits, play an important role in shaping the conversation on the evolution and eventual impact of Industry 4.0 and in preparing citizens adequately to face the challenges created by the fourth industrial revolution. Whether Industry 4.0 turns out to be a net positive or net negative outcome for the world will critically depend on how technology and innovation, as well as the role of machines in society, are closely integrated with human behaviour and humanity.

REFERENCES

1. The ideas and opinions expressed in this discussion paper have been shaped by the author's involvement as a participant in a variety of panels, forums, discussion groups, symposia and other venues around the globe during the past several years. They have also relied heavily on the author's panel participation and presentations at the World Economic Forum in Davos, his conversations with many thought leaders at these events, and his commencement addresses at a number of universities in North America, Europe and Asia, and his lectures at various national and international conferences. These presentations include: panel discussions at the World Economic Forum Annual Meetings in Davos in 2013-2019; commencement speeches delivered at Indian Institute of Technology Madras, Carnegie Mellon University, Indian Institute of Technology Roorkee, Mangalore University, and National Institute of Technology in Tiruchirapplli, India, and at TEDx@CMU and TEDx@NTU during 2017-2019.

2. Author's lectures delivered at: (a) Times Higher Education Summit on 6 February 2018, Shenzhen, China. (b) High Level Dialogue on ASEAN Italy Economic Relations on 11 April 2018, Singapore. (c) ASEAN Conference 2018 on 3 May 2018, Singapore. (d) 31st CIO Workshop on 15 May 2018, Singapore. (e) InSpreneur2.0 on 31 May 2018, Singapore. (f) TedxNTU Talk on 26 August 2018, NTU Singapore. (g) Tradevents Connect 2018 on 30 August 2018, Singapore. (h) Agency for Science, Technology and Research (A*STAR) Leaders in Science Forum on 4 September 2018, Singapore. (i) Lien Development Forum on 7 September 2018, Beijing, China. (j) RWS 2019 Kick-off Seminar on 17 September 2018, Singapore. (k) Deep Tech Summit on 18 September 2018, Singapore. (l) Magee-Womens Research Summit on 9 October 2018, Pittsburgh, PA, US. (m) QS-APPLE Summit on 21 November 2018, Seoul, South Korea. (n) SingHealth Distinguished Lecture on 13 April 2019, Singapore.

3. Klaus Schwab, *The Fourth Industrial Revolution*. The World Economic Forum, Geneva, Switzerland. https://www.weforum.org/about/the-fourth-industrial-revolution-by-klaus-schwab.Seealso:https://www.britannica.com/topic/The-Fourth-Industrial-Revolution-2119734.

CHAPTER 19

The transformative power of the university: the key role of higher education in a sustainable future

Bert van der Zwaan

INTRODUCTION

The university is one of few institutions surviving the changes that have affected society over the past 800 years. Stemming from a period which was dominated by the church and feudal lords, it successfully negotiated the Renaissance and Enlightenment, the industrial revolution of the 19th century, and the profound societal changes following World War II. One of the reasons behind its success is that over most of this time the university was held in high regard, primarily because of the value of its knowledge in combination with its increasingly independent position towards political and religious doctrines. But the success also stemmed from the fact that the university followed the societal mainstream, and avoided biting the hand that fed it. In other words: it was also the result of careful and diplomatic manoeuvring in order to drum up sufficient support and funding from society.

The balance between leading and following, a dilemma the university constantly has been confronted with, is nicely illustrated by the life of Galileo Galilei. He was the founder of the modern natural sciences, and famous already in his time. As such, he was the protégé of the Medici family, and there were many instances in which Galileo needed to operate carefully in order to secure their financial support. It is well documented that the Medicis gave directions and made suggestions for his research. It is equally well documented, however, for instance in his famous letter to the Grand Duchess Christina regarding the heliocentric worldview, that in certain matters he took a completely independent stand.

During many moments in its history, the university has been a leader, pointing the way to uncharted intellectual territory. Most of the technology we consider as normal today stems from curiosity-driven research in the natural sciences, of which the importance was not yet clear at the time it was performed. Equally important has been the contribution of the humanities and social sciences to a new worldview, in which our perception of nature and the world around us fundamentally changed. Think only of the shock induced by the more and more convincing theory of evolution after the introduction of Darwin's first ideas. It had a tremendous impact on theology and philosophy. Think of the extraordinary idea of the universe being 15 billion years old and that now we can still pick up signals from that past. This knowledge created a totally new perception of ourselves as humans — and most of this knowledge was not commissioned or specifically paid for, but the result of blue-sky research carried out by independent scholars. Yet, over the past decades the other side of being a university has become more and more prominent. In particular, since the growth of the university into an institute of mass education, governments could not keep up the level of funding. In the neo-liberal climate of the 1980s, the entrepreneurial university took over, which adopted a business-model partly comparable to industry and became, just like Galileo, more dependent on private funders. Of course, in return for money, these funders took part in the decisions on research priorities, forcing the university into a role of following external agendas.

With the increase of private and competitive funding since the 1980s, universities have become more and more economy-driven. We have seen a seemingly boundless growth of the medical sciences, and to a lesser extent of the technical and natural sciences. Was this the result of legitimate research questions, or is it pushed by industry and society? In other words: how is the research and teaching of a university or a nation prioritized? How much of it is curiosity-driven and to what degree is it based on societal needs? But, perhaps more importantly: is the university passively following these external pressures, or is it making independent decisions based on its own criteria regarding what constitutes valid and urgent research and teaching? Framed in yet another way: is the university an inspirator, independently searching for the best solutions for a sustainable future, or is it simply following the money?

TAKING STOCK: WHERE ARE WE TODAY?

Manuel Castells (2001) defined the role of the university as consisting of four components: the university as ideological apparatus, certainly during its early history closely connected to church and state, the university as mechanism of selection and socialization of dominant elites, the university as generator of knowledge, and, finally, the university as place of training of a skilled workforce. Castells suggested that nowadays the first role of the university is of minor importance only, and that the fourth role is mostly for vocational institutes.

Through history, the ideological relevance of the university has decreased since the Enlightenment. Around that time the conceptualization of absolute freedom of scientific research was a turning point, cutting the ties between the university and state or church. Especially after World War II, this accelerated through secularization and the lifting of many socioeconomic barriers. Simultaneously, the university also lost its position as mechanism to select and socialize elites; instead, it became instrumental in the emancipation of the middle classes, and less and less intended only for the elite, although, in particular, some selective anglophone universities still have this elite-producing function. In the process of massification, the role of the university in training a skilled workforce became more important. But it is the fourth role, the university as generator of knowledge, which has become most prominent. The volume of research has almost exploded over the past 50 years and in research-intensive universities now is even more dominant than teaching.

Of the four roles defined by Castells thus only two are left. The modern university is first and foremost about exploring new knowledge domains, and about training young people to do so: research and education of a skilled workforce form its heart and soul. Simultaneously, with the reduction of the number of roles and developing into institutions of mass education, universities undeniably have become gradually more dependent on outside sources of income. There is evidence of university funding being to a large extent a reflection of the type and state of a nation's economy (Figure 1; Rathenau, 2019a). For instance, in countries with a strong manufacturing industry like Germany, Japan and Korea, the funding of natural sciences and engineering is significantly higher than in countries without such an industry. This shows in an indirect way that the nature of the economy is a prime driver in the priority setting of the research agenda. It follows that the boom in industrialization and advancing technology in the 20th century went hand in hand with the increasing prominence of the natural sciences. Later, with the increasing importance of high tech, the technical sciences gained in importance in particular with the boom of computer sciences.

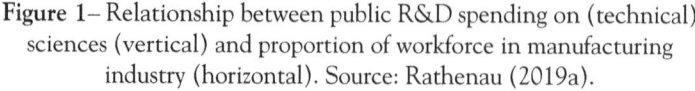

Figure 1– Relationship between public R&D spending on (technical) sciences (vertical) and proportion of workforce in manufacturing industry (horizontal). Source: Rathenau (2019a).

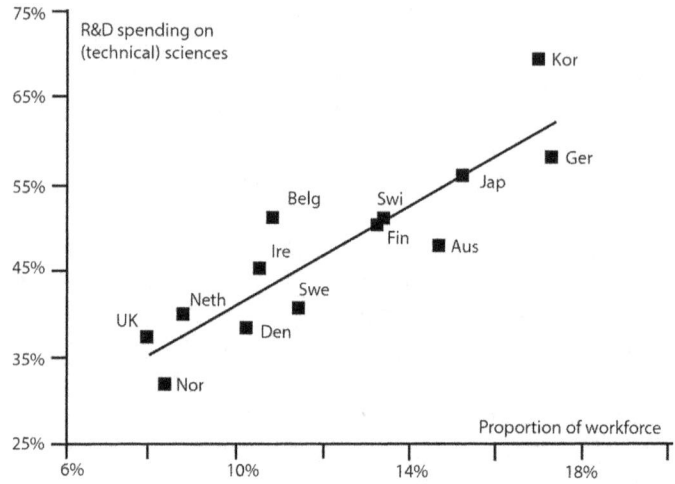

A country like the Netherlands perfectly reflects this international trend. The Dutch economy is a typical knowledge- and service-economy, and the manufacturing industry is no longer dominant. Consequently, investment in the technical and natural sciences is relatively low, like in Norway and the UK, following the relatively low prominence of the manufacturing industry. Superimposed on this basic pattern, over time some substantial shifts in research priority are easily detectable (Figure 2). Over the past 25 years, the biomedical sciences have received a rapidly growing amount of funding, in the US resulting in about a six-fold increase. Only the engineering and computer sciences could follow this trend to some extent, but their funding only doubled. Before the 1980s, the physical and technical sciences were the best funded disciplines; this pattern was presumably already established during World War II under the influence of advancing technology. And, looking even further back in time, we see that the relative importance of the humanities and social sciences was much larger than immediately after the war, and certainly larger than today; over time, the absolute funding of these disciplines shows an almost flat line, which means a relative decrease, since no strong rise in outside funding occurred as in the technical and biomedical sciences.

It is remarkable that the development of the biomedical sciences is totally unrelated to the fundamental economic driver mentioned before. Irrespective of the type of economy, the expenditure for (bio)medical sciences is extremely high in western countries, and still increasing. One could presume that the strong growth of the belief that life is malleable, induced by the tremendous progress made by the biological and technical sciences over the past 25 years, makes "human health" such a strong second driver of the knowledge agenda. Moreover, it seems that the more prosperous a nation is, the more it invests in prolonging life (Figure 3; OECD Health Statistics in Sawyer & Cox, 2018). But it's remarkable that most of this funding is invested in highly technological care for a few, instead of preventive research to safe many. There is more money available for top clinical cancer research than for the prevention of malaria. This suggests that underlying these trends in biomedical sciences, there is a significant impact of the medical-technological industry, which partly did away with their own research labs around 25 years ago and started to collaborate with the universities.

Figure 2– Trends in federal research by discipline USA, 1970-2012. Source: Benjamin *et al.* (2017).

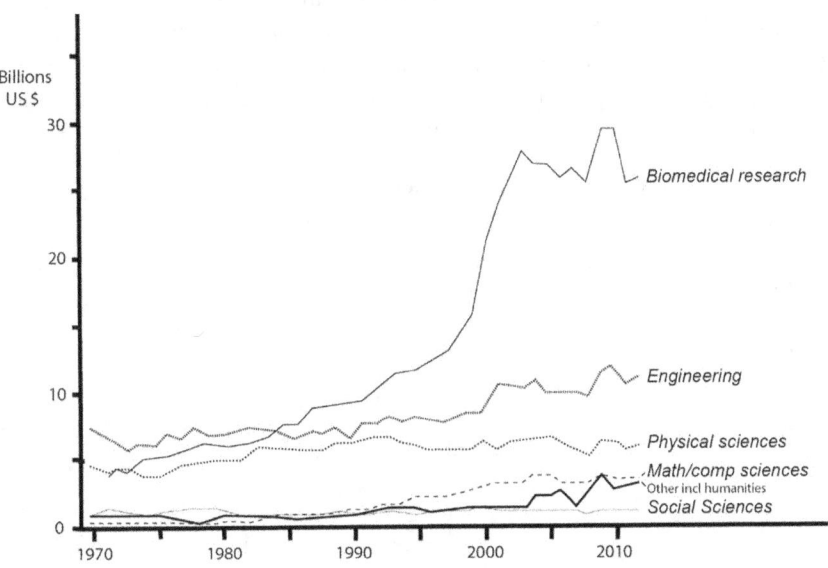

The investment into research in terms of capital is perfectly matched by the capacity generated. In the Netherlands about 70% of the university staff and faculty are working in the medical, technical and natural sciences. The medical sciences alone take a slice of about 30% of all personnel. This, of course, is reflected by the output. The Netherlands belongs to the world's most productive countries in terms of scientific output, but this is even more pronounced in the natural and biomedical sciences: these disciplines produce 35 and 40% respectively of all publications (Rathenau, 2019b).

This rough data seems to implicate that the research agenda of Dutch universities is prioritized in the first place by the nature of the national economy,

and secondly by the international trend of explosive increase of the biomedical sciences with increasing prosperity. Looking at the funding streams in
the Netherlands, there is a slight increase over the past decade in the total
budget for research, but relatively by far the strongest growth is from industry
funding. Over the same period, in particular biomedical sciences increased
in volume, suggesting that these disciplines might have profited most from
the increase in outside funding.

Sometimes, clearly other mechanisms of setting the research agenda are
in place; in many cases this concerns attempts of governments to combine
industry and science policy. A prime example of this is Singapore, which
traditionally has an exceptionally strict science policy almost completely
based on the national technical and innovation priorities. This is not to say
that outside these priority areas no other research is possible, but it signifies
strong steering through earmarking of the funding streams. Another and less
successful example of mixing science and industrial policies is the so-called
top-sector policy of the Dutch government, which started some ten years ago:
this policy in particular stimulated the biomedical, technical and agricultural
disciplines because these were thought to be essential to the Dutch economy.

Figure 3– GDP and health spending per capita,
2017 in US dollars. Source: Sawyer and Cox (2018).

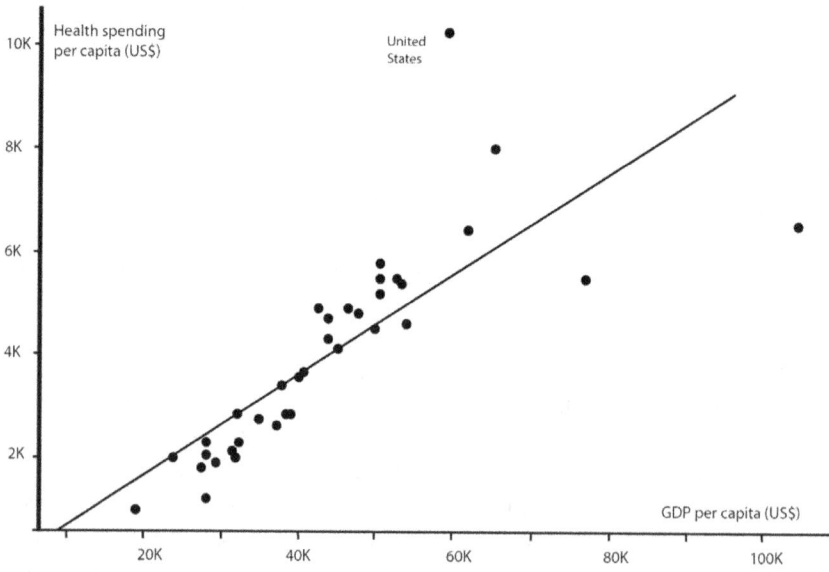

Instead of looking at universities as totally free in setting their research
agenda, the picture clearly is much more conflicted and complex. We already
found that the first layer of influence is formed by the nature of the economy,

and the driving forces ensuing from health care demands. A second layer is formed by attempts to mix industry and science policy, favouring parts of the science spectrum considered to be especially strong in, or beneficial to, a nation. A third layer where the science agenda is set, is formed by the many lobby groups. For instance, given their size and huge output, the medical and natural sciences are important lobby parties in setting the national, or on a European scale even the EU, research agendas. Physics, in particular astrophysics, and chemistry are good examples: in a country like the Netherlands they are extremely well-organized and able to put substantial pressure on the government and funding agencies in order to maintain their traditional high funding. These three layers of agenda-setting turn universities into rather locked-in institutions, in which generating change is extremely difficult.

UNLOCKING THE TRANSFORMATIVE POWER OF THE UNIVERSITY

The question is whether universities are not too passive in following the prevailing funding trends and should be more active to prioritize for instance grand challenges like sustainability or equity, even in spite of their potentially lesser economic relevance in the shorter term. Put in other words: what is the role and relevance of universities in profound societal changes like the ones we are facing today? What is, or what should be, the transformative power of the university? Many universities have impressive missions. Most of these focus on excellence, but often also on the role of the university in educating responsible citizens, or leaders of a future society. One would expect that the research agenda, or at least the educational programs, would be geared towards these missions. In practice, however, the grand societal challenges form a relatively minor part of the research portfolio, or the teaching programs, of any mainstream university.

Sustainability in the widest sense, meaning an economical, ecological and political sustainable world, is a case in point. The urgency to transform society into a sustainable one has increased over the past decades. Obviously, climate change is now generally accepted as a threat to the future of the planet. And, of course, also the research portfolio of most universities has seen a shift towards more research and teaching into this direction. There is almost no university where sustainability is not mentioned. However, if we look at the bare facts, it is surprising how few universities have signed up to the global Sustainable Development Goals of the UN. Or how many teaching programs have no sustainability component. And if universities have a focus on sustainability, it is surprising how little funding is available compared to other disciplines. In the context of this paper, the crucial question is whether the universities are too passive with regard to the content of both their research portfolio and teaching program, in pointing out a course towards a more sustainable future. This in spite of the fact that most students are extremely interested in this, and in most cases would like to see that their university is not a follower but a leader in the debate and in setting examples of a more sustainable style of living.

The rather passive, locked-in modus of the modern university is confirmed by an interesting study of Brennan *et al.* (2004), who compared the role of the university in societal transformations in 15 countries. They conclude that this role is weak in economic and political transformations. In the latter one, in particular the protected space offered by the university permits the "building of the new", as recently has been evident in the student protests in Hong Kong and South Africa. Overall, also the contribution to social transformations seems to be rather weak, the university being a place of reproduction as much as one of transformation. The strongest role the university plays, appears to be in cultural transformations, particularly in terms of opening a door to external ideas and experiences in otherwise closed societies. Brennan *et al.*, but earlier also Van Vught (1993), note that in many transformations, inside and outside the university, faculty resist change, using amongst others the "quality argument", arguing that change would effect the quality of the institution.

It is remarkable in how few cases the university has played a leading role in transformative times. In recent history, only the 1968 student revolts would qualify as an event in which the university was not only a workforce- and knowledge-producer, but also an "ideological apparatus" (cf Castells). The latter could be re-framed in modern terms as cultivating citizenship, a task which is much more palatable to the university than being an ideology machine. Noting how invisible the university has been through history in directly contributing to transforming society, it could be argued that it should shift its focus more from only contributing to the workforce- and knowledge-production, to this task of cultivating responsible citizenship and educating future leaders. Reversely, by doing so, universities would become more visible in society and much more instrumental in solving tomorrow's problems. But this means unlocking the university from its present economy-driven course.

RESTORING THE BALANCE BETWEEN INDIVIDUAL INTEREST AND COLLECTIVE VALUE

Overall, and over the past 40 years in which neo-liberalism has prevailed in western societies, universities have become institutes which understand their societal role more and more as contributing economically, either directly by creating economic value or indirectly by producing a skilled workforce. In terms of Castell's four roles, the roles of the university as place of ideology and of educating the future elite, have been much reduced. Also the knowledge production itself increasingly has been valued in economic terms, illustrated by the present-day emphasis on the economically productive disciplines. Without doubt, the increasing dependency on outside funding has led to changes in the priority-setting of the research agenda. In addition, the individual interest to obtain this funding has become leading, as was the emphasis on rewarding individual performance. This has made the university more a passive follower than a breeding ground for change, or a protected place where ideas for the future are nurtured. In particular, the focus on excellence has stimulated a culture of maximizing output and innovation, focusing on the engineering and natural sciences, and biomedical disciplines, without at the same time stimulating the social sciences and humanities to lead thought formation on social innovation and political ideology. As such, many modern universities tend to be "lopsided", and rather technocratic institutions. Precisely therefore their transformative power is limited.

Many leaders defend this rather technocratic role by stating that the university is not about societal problems, should be neutral, or should not be involved in politics. In the libertarian society of today many would call moral debates dead-ends, in which conflicting personal views would derail the university and disturb its core tasks. Although this is understandable, at the same time these criticasters should realize that taking moral positions is inevitable and forms an inseparable part of our daily university life. This is demonstrated by cases like admissions of minorities, establishing the boundaries of free speech and dealing with hate speech, in cases of integrity, or having patents and earning more money with them than the research subjects, teaching students moral standards, and deciding on divesting or investing in fossil fuels. Sachs (2015) discussed a whole list of such moral problems in which the university is forced to take position. He argues that the university should leave its libertine position in order to take a more moral position, because without morality society disintegrates. Interestingly, he contrasts the prevailing modern American and UK view of morality starting with the protection of the individual from the rest of society, with the view of Aristotle that each individual has the purpose, the *telos*, to mould himself to be a good citizen, a good member of the *polis*. According to Sachs, in order to shift from the first to the latter position each university would need "a framework of guiding principles, and a means of decision-making, that our community should develop and hone in order to answer questions of 'should'."

Instead of the modern rather passive model, one could envisage an active model in which the university not simply follows the funding streams, but actively tries to prioritize based on certain values the university holds high, or moral positions in Sachs' terms. The mission of a university should be the point of departure of a much more rigorous and active strategy in planning teaching and research. If the mission is directed at educating leaders of the future, then it is inevitable that in all teaching programs citizenship and custodianship for a sustainable world are prominently present. This includes also a debate on the public role of universities in querying whether the race for innovation leads to a really sustainable future. As for research, it requires that a prominent place is given to all programs devoted to solutions for the future. This even could imply cross-financing where the underfunded programs are supported with means that are skimmed off the traditionally well-funded disciplines.

Just as teaching and research should reflect the mission of a university, also the campus needs to be in line with this. A sustainable campus should inspire to search for ways to a more sustainable world: it is clear that maintaining our present western style of life is no option. Our lifestyle needs to be restructured drastically and campus life should lead the way. Innovations that are not contributing to this should have no place, whereas innovations providing solutions for the future should be embraced. Instead of being a place where history dominates, campuses should be breeding grounds of innovations and training for another life that is in line with a much reduced ecological footprint.

Modern university leadership is to a large extent consumed by stimulating and maintaining excellence in teaching and research, and secondly by obtaining funding from a large variety of sources. As such, and certainly if the wishes of the faculty are followed, leadership strategy could rapidly become reduced to a strategy of "follow the money". It requires strong leadership to change this pattern and to play a role in the societal transformations ahead of us. In particular contributing to the cultural changes requires active agenda-setting and creating a strong awareness among the faculty of what in this context the university's mission is. Instead of being a follower, the university should be more an inspirator and leader. To realize this, the university needs to focus differently, not only on educating a skilled workforce and furthering knowledge, but also on creating a new elite, a generation of leaders with great awareness of the grand challenges ahead of us. But maybe most of all, to unlock the university requires restoring the balance: from a university driven by individual interests and rewarding of individual performance, to one with a more collective, value-driven viewpoint of what social, political and economic sustainability means for the next coming decades.

ACKNOWLEDGEMENTS

I thank the organizers of the Glion Colloquium 2019, which was inspirational, and Sijbolt Noorda and Frans van Vught for stimulating discussions.

REFERENCES

Benjamin, G. S., Brown, L. & Carlin, E. (2017) *Strengthening the disaster resilience of the academic biomedical research community*. National Academy Sciences.

Brennan, J., King, R. & Lebeau, Y. (2004). *The Role of Universities in the Transformation of Societies*. Centre for Higher Education Research and Information, London.

Castells, M. (2001). *Universities as dynamic systems of contradictory functions*. African Minds Publishing, Cape Town.

Rathenau. (2019a). Rathenau Instituut. "Development of the scientific research profile of the Netherlands". Rathenau Factsheet (in Dutch), The Hague.

Rathenau. (2019b). Rathenau Instituut. "R&D expenditure and capacity by field of science." *Rathenau Factsheet*. The Hague.

Sachs, J. D. (2015). "What is a moral university in the 21st century?" Speech at Columbia University, 30 March 2015.

Sawyer, B. & Cox, C. (2018). "How does health spending in the U.S. compare to other countries?" Kaiser Family Foundation, San Francisco.

Van Vught, F. A. (1993). *Governmental Strategies and Innovation in Higher Education*. Jessica Kingsley Publishers, London.

CHAPTER 20

The Three (Four) Pillars of Sustainable Development or "The Great Race"

Timothy Killeen

I recall a slapstick film from back in the mid-1960s with the title The Great Race. In it, the quintessential hero (the Great Leslie, dressed in white, of course) is challenged by a despicable and traditionally melodramatic villain known as Professor Fate, who proposes an epic over-ground automobile race from New York to Paris, travelling the long way across Siberia. Despite a massive pie fight, promoted at the time as the biggest one ever, and Fate's many scurrilous attempts to cheat along the way, things work out in the end, although not without extensive damage to the iconic Eiffel Tower!

The title of the movie — as well as some of the movie's intense drama and confusion — came to mind as I was thinking about the subtopic at hand: the three pillars of sustainable societal, ecological and economic development. Let me explain.

We do indeed face a momentous race between two competing, fast-developing and, at times, countervailing tendencies. The first is the acquisition of sophisticated knowledge about the complex and non-linear relationship between humankind and the planet that supports and nurtures all life. The second is the absolutely urgent need for innovative technologies to be deployed to improve human welfare and, at times, to avert catastrophes. It is abundantly clear that we need more "deployable innovation for sustainability" — and need it now.

In many ways, this "great race" informs the work of our university system, because it is "on our watch" that this race needs to be won. If we would have had the sophisticated current-day biophysical and chemical understanding 100 years ago, then many of the "wicked" problems we now face — e.g. resource scarcity, biodiversity loss, poor air quality, deleterious climate change and its severe weather impacts, fresh water unavailability, food and soil degradation, and conflict avoidance — would, quite possibly, have been long ago resolved. Conversely, if today's deepening knowledge were still 100 years off into the future, then we would, in all likelihood, have no chance of avoiding ecological and societal collapse. Sometimes, it seems to me to be a coincidence of cosmic proportions that the required knowledge is emerging at the very time that humanity needs it. On our watch.

So, what is the role of a large public university system in this, the "great race" of our times? As president of the University of Illinois System, I think about this often. Our system has nearly 86,000 talented students enrolled in three universities across the state of Illinois, more than 750,000 living alumni, and roughly $1 billion per year in externally funded research, with faculty expertise covering most if not all fields of intellectual interest. It also has a formal and deeply felt mission to serve the public good through its original land-grant university in Urbana-Champaign, its large research-intensive public university in Chicago (the third largest city in the United States), and its comprehensive liberal arts university in Springfield, the state capital. Each university has a distinctive character and setting, and a different range of focus. For example, the University of Illinois at Chicago is home to one of the nation's largest medical schools and an expansive, innovative healthcare system focused on population medicine in a world city. The University of Illinois at Urbana-Champaign has a highly ranked engineering school with special renown across the computer and information sciences. And the University of Illinois at Springfield has particular expertise in public policy, criminal justice and Abraham Lincoln studies. Despite these very complementary differences, all three of our universities share in a common mission — to serve the public good.

A university system blessed with our assets must, then, drive the rapid development of new knowledge and technologies that can be deployed to build and sustain human prosperity. We intend to work on this as individual universities and in the collective, but primarily through extensive partnerships — with governmental, non-governmental and private (commercial) enterprises and individuals.

In our published strategic framework that guides our work, adopted in 2016, we use the terminology: "optimizing impact for the public good." When I think about this kind of optimization, I often use the following simple heuristic equation:

$$II = (EE \times SS)^{MM}$$

Here, I is "impact", which is the element to be optimized. Impact is dependent on both "excellence", E, and "scale", S. Without excellence, it is very difficult to innovate rapidly, and without larger scales, the products of the innovation cannot be deployed as efficiently, either by individuals or through commercialization strategies. This heuristic relationship leads one to a greater appreciation of the impact that a large and excellent public university system, such as ours, can have. In this thinking, 86,000 students carry with them a much larger potential for impact than do a few thousand students, even those from first-rate universities — as long as institutional excellence is not diluted or traded away as size grows. In this equation, the product of excellence and scale is then raised to the power of what I refer to as institutional *Magic* (M). If M is less than unity, the resultant impact is degraded. If M is much greater than unity, then exciting non-linear enhancements to impact happen.

What is the magic? The nominal exponent, M, is essentially here to represent institutional *culture* — all those special things that combine to characterize a vibrant institution. These are elements such as a deep commitment to teaching and learning; visionary and trusted leadership; talent acquisition, recognition and support; collaborative impulses; the ability to build teams and to generate and sustain effective and authentic partnerships; access to major facilities and resources; the ability to navigate and interconnect disciplines; the fulsome embrace of diversity in all forms (approach, background, discipline, etc.); and the willingness to take risks in pushing the envelope of new knowledge. I am sure any reader would be able to develop his or her own list of such attributes. But, with this thinking in mind, those institutions with both scale and excellence that also have a vibrant (i.e. magical) institutional culture can have a tremendous impact on the world.

What, then, is the role of a large, excellent, vibrant university system in building the societal, ecological and economic underpinnings for a sustainable future? I postulate here that such institutions provide the very best opportunities for solutions that can serve society into the future. Going even further, I suggest that these are perhaps the *only* institutions capable of taking on the challenge to win the great race of our times. Even the largest, best-endowed companies can lack the required multi-disciplinary expertise, the central role in developing human capital, and the risk-taking culture. It follows that we, in the leadership of large, public, research-focused university systems, should recognize a special responsibility to act with urgency to solve the grand challenges related to sustainability.

In the next few paragraphs, I provide modest comments on some of the particular approaches that I believe will be essential to success (and add a pillar to the discussion):

EDUCATION

The first imperative (and the fourth pillar!), of course, is the fundamental commitment to lifelong education. It is critically important to have institutions, particularly at the higher education levels, that nurture students' abilities to think critically, to write sensibly and cogently, to exhibit discernment in recognizing what is true and what is false, and to rely on evidence-based decision-making whenever possible. Modern pedagogical approaches should focus on effective and demonstrable learning, teamwork, skills development and a combination of both analytical and critical thinking. In this regard, the social sciences and the arts and humanities are every bit as vital as the canonical science, technology, engineering and mathematics (STEM) disciplines.

I feel it necessary here to single out the scholarly work and education in the social sciences, arts and humanities. As I wrote recently when initiating a system-wide initiative to celebrate the arts and humanities: "Research and creative breakthroughs in these arenas help us imagine new approaches to today's societal challenges, drawing from deep historical experience, finely honed craft, and expertise in collaboration and improvization. The humanities and the arts also serve diverse publics by nurturing the human spirit, by offering inspirational new experiences, renewed connection to records of the past, and frameworks for living within difference and debate."

Although some economic headwinds have undoubtedly harmed the arts and humanities at many universities due to public misperceptions of lower-paid employment opportunities for graduates, I believe that it is very important for university systems like ours to continue to build and support these fields of scholarship for all the richness and benefits they bring to society, including the kind of lateral thinking and problem-solving needed to win the great race.

A last comment here about the social sciences, arts and humanities. When I was the Assistant Director for the Geosciences at the U.S. National Science Foundation from 2010-14, we toyed with avoiding the word "sustainability" and replacing it with "thrivability". Although a bit of a mouthful, the latter term implies that we seek a healthy and secure future for our children — not just one that sustains an imperfect, and perhaps miserable, status quo. We will absolutely need university-based scholarship in the social sciences, arts and humanities — as well as all the biophysical sciences and engineering — to approach a future where the human condition is celebrated and nurtured and humankind actually does thrive.

SOCIETAL PILLAR

There are many challenges associated with sustainability that lie within the province of research universities. Alan Leshner, the long-term former CEO of the American Association for the Advancement of Science (AAAS), described what he saw as the major global societal issues facing humanity in a 2011 talk on the challenges of building a global science community. His list included the following: sustainability; renewable energy; information and communication technology; universal access to education; poverty and economic opportunity; technology-based manufacturing and jobs; intellectual property rights; terrorism and security; disasters; vaccines and medical therapies; quality and accessibility of health care.

It is noteworthy that every one of these issues is under intensive study within universities like ours, with faculty experts engaged from within and across many different disciplines who also are connected to external partners inside and outside government. These disciplines include all of the sciences and engineering, but also the social and behavioural sciences where human decision-making under conditions of risk and uncertainty is a new emphasis. Since such decision-making will be at the very core of successfully addressing the societal grand challenges of our times, the contributions of these non-STEM fields (including economics) will be immeasurable.

While it is very difficult to forecast with any kind of precision the transformative breakthroughs in non-STEM areas that can address these grand challenges directly, it is hard to imagine substantial progress in any of these areas without universities playing a catalytic, central role. Dr Leshner's list interestingly includes "intellectual property rights" — and I take this, in part, as a signal of the growing importance of the kind of public-private (university-industry) partnerships discussed below.

ECOLOGICAL PILLAR

The ecological pillar for sustainable development is, I believe, the most important one. After all, nothing else much matters if the natural platforms supporting human existence erode away from us. The current knowledge base of the state, pressure/response, and resultant changes to the ecological system has been developed — and must be extended and maintained — by means of a healthy university research and development base. A quick look at the authorship and citation listings for the influential and authoritative reports of the Intergovernmental Panel for Climate Change (IPCC) will quickly demonstrate the significance of university-based or university-connected researchers in developing the modern scientific understanding of the human/planet relationship.

Earth system models — using supercomputing technology, and involving many scientific experts worldwide — are quickly improving and now include most of the important coupled ocean, atmosphere, soil and land processes that control the climate system at a high level of sophistication. The Community

Earth System Model (CESM) community model, for example, developed by the National Center for Atmospheric Research (NCAR), where I was director for eight years, has shown an exciting level of predictive skill at both the regional level and over many different temporal and spatial domains. Outputs from this sophisticated class of model — and further developments — are critical to improving detailed knowledge and understanding of what lies ahead of us, contingent on the socioeconomic scenario that society will follow. The NCAR-CESM and other similar models are among the most important human artifacts of our time and will need to be nourished through the continued upgrading of computational capabilities and access to "big data" describing the earth system for scientific validation. It is a continuing triumph of modern science that these complex modeling systems and their outputs are generally available to the public for free, and that future developments continue to be carefully validated in an open-source environment.

In addition to the numerical models, large observational systems are coming of age around the world. Oceanic observatories, ecological networks, seismological arrays and atmospheric remote sensing systems from ground and space are all contributing to winning the great race. An analysis of the National Science Foundation budget will quickly demonstrate how important these large-science infrastructural facilities are to the expert scientific community.

But there are also significant political challenges in further developing and refining this knowledge base and turning it into an action agenda. I recall helping draft the first position statement on climate change and greenhouse gases published by the American Geophysical Union (AGU) in 1999. AGU is the largest professional society of geoscientists in the world (I was later to become AGU president for a two-year term). This first statement has been replaced several times by more comprehensive ones, but I vividly recall the splash that was made in 1999 on its release — at a standing-room-only National Press Club event in Washington, D.C. I was one of a handful of scientists defending the new position statement in the context of the very active and highly charged US presidential election process underway in 1999. I felt very inadequately prepared for the political backlash. The reporters were mainly focused on the *triple negative* phrase in the 1999 report: "*AGU believes that the present level of scientific uncertainty does not justify inaction in the mitigation of human-induced climate change and/or adaptation to it.*" This formulation frustrated many of the attending journalists who wanted greater clarity in terms of an action agenda. Our cautious but scientifically defensible statement, however, was absolutely appropriate for its time, but I confess to a determination to never again employ a triple negative in such work!

Even by 1999, of course, the jury had largely come in on the scientific case for human-induced climate change and the slow-moving but now accelerating threat it was bringing to society.

Unfortunately, the political response to this situation remains muted and insufficient, even 20 years later. Many members of the general public, particularly in the United States, have become convinced that anthropogenic climate change is not real and, therefore, is not something that requires resources to address. I attribute this, in part, to entrenched commercial interests and their effective communication strategies, but also to the fairly muddled presentation of the "kitchen table" implications of the mainstream scientific consensus by the expert community. Once again, future university research — ranging well into the economics, communications, journalism, and public policy domains — will be needed to clarify societal options using our best and most sophisticated quantitative analyses and predictions of change.

ECONOMIC PILLAR

As in all forms of human activity, economic forces will determine the pace and results of societal change related to the new external pressures. Perhaps the first thing to note here is that there needs to be significantly more effort expended on the full-cost accounting and economic impact of changes and pressures. A discussion of carbon taxation is just the tip of the iceberg of what is needed. We will have to develop new *figures of merit*, beyond the dollar, to make and sustain resource allocation decisions. Human welfare impacts need to be quantified and given much higher weighting in such decisions than is the case at present. Key questions abound. What is the true cost of degraded air quality in the GDP of a country and who bears those costs if the polluted air is travelling from elsewhere? What will climate change-induced reductions in crop productivity do for childhood malnutrition and how much will it cost to remediate those effects? What coastal regions should be armored to combat sea level rise and what happens to the insurance costs in other, lower priority settings? Questions like these can and will be answered rigorously and authoritatively in university settings, but that work must commence and be fully funded and energized.

Secondly, it should be realized that there is simply not currently enough funding from all of the world's national science agencies combined to appropriately support the needed research and innovation for sustainability going into the next decade.

So, we must ask the question, how will all this be funded?

Several years ago, I estimated the international level of governmental (funding agency) support for climate science, including all the remote-sensing satellite assets in space today, to be on the order of magnitude of $10 billion per year. Although this may seem like a large investment, it is dwarfed by the costs incurred annually by extreme events such as droughts, floods and heat waves that are all increasing in frequency and severity. In my opinion, the desperately needed augmented funding base for the applied research needs in earth system science will have to come, therefore, from the most heavily affected private sector — notably the finance, insurance and reinsurance houses that underwrite the large infrastructural investments around the world and which are very focused on systemic risk mitigation to control their costs. Such sources of support can and should augment the worldwide research and capacity building base for this kind of research by an order of magnitude into the next decade. These new dollars should be spent, in significant part, in the appropriate university communities.

SUMMARY

Universities should reinforce and augment the bio- and geo-physical research efforts, including all fields of engineering and the critical behavioural sciences. Deployable technological advances and commercialization strategies must be generated rapidly in support of tomorrow's decision-makers. A major (order of magnitude at least) increase in funding levels is needed and this will require tapping into the most heavily affected private sectors.

A recommitment to the educational process to develop the human capital needed for "thrivability" is needed. The deleterious changes associated with climate, air quality, fresh water availability, food production and the like will undoubtedly dominate the narrative of the rest of the 21st century and beyond. Our future students will be extremely motivated to contribute to solutions and will want to be fully prepared to address this complex set of interrelated challenges. In this educational transformation, the role of the arts, humanities and social sciences will all need to be fully integrated.

More and more sophisticated earth systems models with regional fidelity will be required to support important and costly decisions on mitigation, tactical withdrawal, and resource allocation. Universities will need to address not just the likelihood of projected changes, but also the more complex questions of societal adjustment, cost and systemic risk mitigation — terms that more fully resonate with the private sector. In this regard, a public-private-governmental triad needs to be established to create the economic circumstances and partnerships that naturally favour more sustainable activity.

Finally, we will have to invent and deploy mechanisms to decarbonize the atmosphere. As I write this, the carbon dioxide level in the atmosphere has breached 415 parts per million (May 2019) — a level that would have seemed to have been very unlikely and extremely problematic 20 years ago. Active strategies to physically remove greenhouse gases from the atmosphere will need to be designed, developed and piloted. Examples in our own university system include the development of "artificial leaf" technology, designed to remove carbon dioxide from the atmosphere and the testing of large-scale soil additives to enhance weathering processes in agricultural settings. Many other technologies will be needed, involving what is commonly called "geoengineering."

So, the future will be one of extensive public-private-governmental collaboration and partnerships — led and catalysed by universities, with new sources of funding, new and intellectually rich research pathways, and new quality metrics and figures of merit that do not currently exist.

This is what is needed to win the great race.

BY WAY OF CONCLUSION

The University at
the Crossroads
to a Sustainable Future

Much like the 11th Glion Colloquium in 2017, during the 12th Colloquium there was less emphasis on the themes that are familiar among university leaders like financial sustainability, research opportunities, rankings and internationalization. To use a phrase from the concluding remarks of the 11th Colloquium, the "long shadow" of all the political events in Europe and the United States since 2016 "that had been cast across the world" was still palpable. Of course, also during this colloquium the contributions addressed themes that are important and urgent to universities, but throughout the discussions there was a clear sense of the rapidly changing world around us and the question of how universities could adapt to the new reality. It seems that the world as a global village is being replaced by separate political blocks that are fighting trade wars, and in which the days of growing student mobility and increasing internationalization are over.

Reflecting on all the contributions to this colloquium, it appeared that they could be grouped into three clusters dealing with the Global, the Local and the Future, respectively. Following this subdivision, going from the global to the future, we summarize below some of our thoughts on the rapidly changing context in which universities probably need to operate over the coming years, followed by a number of suggestions how universities could collaborate successfully on global and local levels in order to face the challenges of the future.

HOW HOPE FADED

The final two decades of the previous millennium showed a staggering change and ended full of hope. Worldwide political relationships including those between the superpowers were less strained than ever before, and the Iron Curtain ceased to exist. The idea that the world is a global village rapidly became true due to a surprising acceleration of mobility and connectivity. The mobile phone, a novelty in the 1980s, gained ground with incredible speed. The most remote places on earth became connected and, through that, part of the world's events. The even more surprising development of IT opened up a completely new world, which in this millennium continues to surprise.

Due to all these developments, the university, about 800 years old at the end of the previous millennium, received new momentum, not in the last place from the revolutionizing impact of digitization. In addition, student mobility increased, eventually leading to massive streams of students, in particular from Asia studying in the West. Internationalization led to an unprecedented exchange between scholars and scientists from all over the world. Looking back, these decades were almost like a new Renaissance, showing the birth of a global knowledge ecosystem in which digitization was as important as printing had been 600 years earlier. Rightfully, this period is now considered to be the beginning of the fourth industrial revolution.

How strong is the contrast between these final decades of the previous millennium and the first two of the present one! This millennium started with the launch of the first university rankings, enhancing the competition between universities and potentially threatening cooperation. It also became evident that mass education, at first glance a resounding success of universities since the 1970s, had led to the worldwide rise of a middle class that more and more diverged from a lower class that felt left behind. This became painfully clear during the second decade of this millennium, in the aftermath of the financial crisis of 2008-2011. The lower class in particular was seriously affected whereas it turned out that the higher educated part of the population still had more opportunities on the global labour market. This divide between haves and have-nots coinciding with the level of education is patently clear in the conditions in which people are forced to live. Today, even in a rich country like the US, the average regional difference between minimum and maximum life expectancies is increasing. It is only recently that we started to realize that this growing social gap is one of the fundamental reasons behind the polarized political landscape, especially in Europe and North America.

The financial crisis also revealed one weakness of globalization in the sense that a crisis in one part of the world is more rapidly felt elsewhere. As a reaction, protectionist and populist views surfaced and rapidly became mainstream in politics. Whereas at the end of the previous millennium leaders all over the world had pledged to open political and trading systems, the trade war between the US on the one hand, and Europe and China on the other, which started at the end of 2018, tells an altogether different tale and shows how profoundly the world has changed.

So far, all these changes can be regarded as part of the movement of a pendulum, or the result of action and reaction, in the sense that we can be hopeful that conditions rapidly restore to "normal". Because, in spite of the negative developments, one still has to conclude that over the past decades the world has become a better place, in which the state of welfare is higher than ever, and safety has increased for many, whereas overall violence decreased.

However, the second decade of our millennium showed one strongly disturbing sign that threatens to take all hope away. It was for the first time since World War II that we saw such massive migrant streams: 70.7 million people were forcibly displaced worldwide in 2019, while 25.9 million people were living as refugees. Part of this is the result of conflicts, but what is frightening is that more and more migrant streams are induced by climate change and natural disasters. In 2018 the UN General Assembly almost unanimously recognized that "climate, environmental degradation and natural disasters increasingly interact with the drivers of refugee movement". The availability of water as an elementary resource is endangered, and the number of hot spots which are too hot and too dry, or too cold and wet, is rapidly increasing. The conclusion seems unavoidable that the unprecedented improvements in health care and the connected growth of the world's population, in combination with equally unprecedented technological development, have led to a situation in which we have reached the boundaries of the system.

In itself this already poses a huge challenge to the world community. But matters are compounded by the lack of effective leadership that the superpowers, or for that matter national governments, display. Where, in the aftermath of World War II, leaders took effective steps to enhance collaboration and forge world leadership in the context of the new "united nations", we now observe the disintegration of global leadership and decreasing effectiveness of national leadership. It is as if traditional leadership as we have experienced it over the past millennium is not as effective in the new one. The superpowers are weakened and not individually able to take the lead or settle issues. There are increasingly louder suggestions that national governments are failing to deal with the new challenges, and that we have entered a phase that is characterized by a fundamental questioning of multilateralism. Instead, nations are more and more focusing on their own interests. It is clear that the world is waiting for new groupings of decision-makers, able to cut across national interests and act on the global level needed to face global challenges.

AT THE CROSSROADS TO A SUSTAINABLE FUTURE

Universities are among the institutions that are able to transcend national boundaries and interests, and are, by their very nature, multilateralist. Without making concessions to excellence and independence, universities could take leadership by forming consortia, collaborating with other universities, industry and other parties, like cities and regions. Working together, they could formulate universal goals in line with the United Nations Development Goals, and, by collaborating with industries, cities and regions, they can translate this collaboration into regional impact. However, this requires bold steps and demands a new type of leadership that is not afraid to change course in order to give the university the central place in society it deserves, but, most of all, by doing so gives new hope to young people for a better custodianship of this world and its future generations.

Reflecting on all the contributions to the colloquium, we think that universities could and should play an active and visible role in laying the foundation for a sustainable society. They could do that by pursuing the following goals:

1. Preparing young people for the future

In the first Glion Declaration (1998) it was clearly stated that "teaching is a moral vocation, involving not just the transfer of technical information, however sophisticated, but also the balanced development of the whole person". Therefore, in addition to transferring knowledge, the emphasis of teaching should also be on cultivating a keen eye for the needs of society, developing a feeling of responsibility for the future, and the development of ethical norms of what is desirable in view of a more equitable society in which resources are fairly shared. Universities should actively prepare their students for the future and impress on them the need for leadership and responsibility, which follow from the privilege of having enjoyed higher education.

2. Being a laboratory for new leadership

Classically, the core task of a university is the custodianship of knowledge in the widest sense. By nature, this involves training young people for their future roles. Often greatest emphasis is on scientific training, whereas the formational aspects are overlooked. As much as about knowledge, university training is about crafting a lifestyle characterized by the ability to identify and solve problems, to ask relevant questions and question problematic reasoning. At the same time, training for the future also involves teaching how to keep open a keen eye for what is going on in society and what responsible citizenship entails. More than ever before, students should be trained for leadership that involves all of the above-mentioned qualities.

This cannot be done really successfully in traditional education, involving a rather passive role for the students listening to the teacher. This calls for challenged-based or problem-driven teaching, during which "soft skills" can be transferred much more effectively.

3. Providing relevant knowledge to society

It is essential that universities are autonomous and independent in setting their research and teaching agendas. However, that does not imply that the university is an ivory tower. On the contrary, the university should recognize its social responsibility by delivering knowledge that is essential for the solution of the problems we are facing today and in the near future. The university should do so impartially, positioning itself on the crossroads of fundamental research and large societal problems, and teaching students in the same vein. The agenda setting of the university should preferably take into account the United Nations Sustainable Development Goals.

4. Contributing to bridging the social divide

Universities should be aware of the societal gap that is growing rapidly, and which to a large extent is connected to having enjoyed higher education or not. The higher educated have much better opportunities in the global labour market and look towards the future with confidence. The flipside of this is that worldwide the less educated part of the population is lagging behind because of their lesser potential in the rapidly changing labour market and they are therefore more vulnerable to the negative effects of economic change. The responsibility of being an institute of higher education is not only to take care of one's own students, but also to reach out to the less educated parts of society. Lifelong Learning is essential in this respect, a task in which universities could and should play a pivotal role.

5. Setting the example for a sustainable future

To be convincing in assuming leadership and custodianship in the widest sense, it is essential that universities create an environment which reflects this. In other words: universities should put the money where their mouth is. Where possible, the university campus should reflect the ambition to create a sustainable environment in terms of saving energy and greening the campus in a variety of ways. And debates on seemingly lesser questions like (not) serving meat in the university restaurant are valuable experiences as a university community. This is all because the campus and campus life should not only be about what the present generation of university leaders and students find nice, comfortable and pleasant, but as much about letting students experience lifestyles which are fit for a future in which a much more modest and sharing lifestyle is imperative.

Furthermore, we think that university leaders are uniquely placed to give the university a key position in society by realizing the following:

6. Taking the lead

In addition to research and teaching, since the 1990s universities have focused on services to society. In this ongoing process of stepping out of the ivory tower, universities should strive to lead in a world that evidently is trying to find new leadership structures. Where traditional governmental leadership is failing, new groupings take the lead or meet at, for instance, the World Economic Forum. Similarly, networks of large cities or consortia of regions try to shift the balance of power to their advantage. This involves more than lobbying: more and more it entails taking steps towards securing a sustainable and prosperous future where national governments fail to take such steps. It is essential that universities take the lead in this process, not only in the area of higher education, but also in a wider sense as institutions that can bring knowledge and wisdom to the debate. In order to be visible and be heard, universities should join forces and consciously develop a strategy of investing in an agenda of collaboration more than investing in rankings and competition.

7. Being bold and visible

University leaders are charged with the duty to keep an extremely diverse community together, a community that is, moreover, composed of highly individualistic thinkers. Serving this heterogenous community often means that boldness or outspokenness pleases one part, but antagonizes another. Therefore, university leaders are by nature careful and relatively conservative. However, what these times call for is boldness in the vision that universities should lead and be visible, in spite of the possible protests from established scholars claiming that it is only "the quality of research" that counts. In this context, it is crucial to listen to the voices of the students and younger members of the scientific community: it is their future which is at stake.

8. Strengthening international university networks

Universities are used to collaborating in an international context: multilateralism is at the very foundation of free exchange of ideas and scientific progress. However, most university networks are now focusing on lobbying for funding and position, sometimes also on improving research and teaching.

What is needed, however, is a concerted effort of universities to collaborate for a better future aiming at, for instance, the realization of the United Nations Sustainable Development Goals. Together with partners from industry and NGOs, universities could be powerful players in turning the tide of challenges and contribute together in using knowledge to solve the problems of the future.

9. Implementing a decision-making support system

It essential that international organizations are connected to the latest knowledge, technology and evidence as produced by scientific institutions. To this end, universities could form hubs of multilateral science diplomacy, because only global networks of leading research institutions can harness the breadth of interdisciplinary evidence, knowledge and perspectives that are needed to tackle the complex global issues and multifaceted societal challenges. Moreover, it is only through highly visible collaborations like these that sufficient players from the private and philanthropic sector can be engaged to make the necessary impact.

10. Becoming once again a place for hope

In spite of the many disturbing developments, we are still living in a time in which unprecedented steps are taken in gaining more prosperity for many. In spite of the numerous local conflicts and political tensions, we are still living in times with unprecedented low levels of violence. In spite of the huge challenges ahead of us, we are still living in times when knowledge and wisdom can make the difference. In addition to being places for training young people for the future, universities can be places where young people are also filled with hope and idealism, which are much more effective weapons to fight the demons of the future than anything else.

Prof. Luc Weber   Prof. Bert Van der Zwaan
Rector Emeritus University of Geneva   Rector Emeritus University of Utrecht

Prof. Yves Flückiger
Rector University of Geneva
President The Glion Colloquium

www.ingramcontent.com/pod-product-compliance
Lightning Source LLC
Chambersburg PA
CBHW070325220526
45467CB00001B/34